SpringerBriefs in Physics

SpringerBriefs in Physics are a series of slim high-quality publications encompassing the entire spectrum of physics. Manuscripts for SpringerBriefs in Physics will be evaluated by Springer and by members of the Editorial Board. Proposals and other communication should be sent to your Publishing Editors at Springer.

Featuring compact volumes of 50 to 125 pages (approximately 20,000–45,000 words), Briefs are shorter than a conventional book but longer than a journal article. Thus, Briefs serve as timely, concise tools for students, researchers, and professionals.

Typical texts for publication might include:

- A snapshot review of the current state of a hot or emerging field
- A concise introduction to core concepts that students must understand in order to make independent contributions
- An extended research report giving more details and discussion than is possible in a conventional journal article
- A manual describing underlying principles and best practices for an experimental technique
- An essay exploring new ideas within physics, related philosophical issues, or broader topics such as science and society

Briefs allow authors to present their ideas and readers to absorb them with minimal time investment. Briefs will be published as part of Springer's eBook collection, with millions of users worldwide. In addition, they will be available, just like other books, for individual print and electronic purchase. Briefs are characterized by fast, global electronic dissemination, straightforward publishing agreements, easy-to-use manuscript preparation and formatting guidelines, and expedited production schedules. We aim for publication 8–12 weeks after acceptance.

More information about this series at http://www.springer.com/series/8902

Michele Campisi

Lectures on the Mechanical Foundations of Thermodynamics

 Springer

Michele Campisi
Istituto Nanoscienze
Consiglio Nazionale delle Ricerche
Pisa, Italy

ISSN 2191-5423   ISSN 2191-5431   (electronic)
SpringerBriefs in Physics
ISBN 978-3-030-87162-8   ISBN 978-3-030-87163-5   (eBook)
https://doi.org/10.1007/978-3-030-87163-5

This Springer imprint is published by the registered company Springer Nature Switzerland AG
The registered company address is: Gewerbestrasse 11, 6330 Cham, Switzerland

To Elisa.

My life.

To the memory of Prof. Donald H. Kobe.

Mentor and example.

Foreword

Thermodynamics, conceived more than 150 years ago by Carnot, Clausius and others, has been experiencing a remarkable resurgence over the last two decades, spurred by major experimental and technological advances that enable the direct observation and manipulation of quantum gases, DNA molecules and other nanoscale systems. Originally developed to describe heat exchange and work extraction processes at the macroscopic scales, it has become increasingly clear in recent years that thermodynamic concepts and principles provide an equally powerful framework for the theoretical description of energy and information transfer in microscopic systems, from quantum heat engines to biological cells. Michele Campisi, whom I first met while still a doctoral student in Peter Hänggi's group at Augsburg, has been one of the pioneers at the forefront of this rapidly expanding research field.

In his monograph *Lectures on the Mechanical Foundations of Thermodynamics*, which is based on courses he taught at the University of Florence, Michele discusses how a thermodynamic framework can be systematically constructed from Hamiltonian dynamics. This fundamental question has a long and interesting history, dating back to seminal work by Boltzmann, Gibbs and Hertz at the turn of the twentieth century, and it has regained critical importance with recent applications of thermostatistical concepts to small systems. While many modern textbooks adopt an axiomatic approach that starts from an abstract set of thermodynamic postulates, the constructive 'bottom-up' approach pursued in this monograph shows how a reduced thermodynamic description emerges naturally by linking dynamical time averages with statistical ensemble averages through the ergodicity hypothesis.

Rather than aiming for a broad scope of applications, Michele's lectures focus on building a deep and succinct understanding of thermostatistical core concepts. Carefully chosen pedagogical examples highlight the different physical assumptions underlying the microcanonical and canonical ensembles. Ensemble equivalence, often taken for granted in the thermodynamic limit, can be and is violated in small (and many not-so-small) systems, which has profound implications for the accurate description of energy conversion and transfer in microbiological and quantum processes. Students and researchers interested in developing a strong foundational understanding of thermodynamic concepts will greatly appreciate the guidance provided by this monograph.

Over the last fifteen-plus years, I have learned immensely from my many discussions with Michele, as well as from studying his original research papers and review articles—after reading his monograph, you will surely feel the same way.

Cambridge, MA, USA   Jörn Dunkel
August 2021

Preface

The largest majority of textbooks presents the subject of statistical mechanics by adding one or more postulates to the microscopic theory that describes the dynamics of the constituents of matter, be it classical mechanics or quantum mechanics. Typically one postulates the entropy and then derives thermodynamics therefrom. However, that is not desirable from a foundational point of view, because when adding a postulate to an existing theory, there is always a risk that it is not fully coherent with it.

The present series of lectures provides an introduction to statistical mechanics, specifically the theory of statistical ensembles, that does not require any further postulate besides the laws that govern the Hamiltonian (classical) dynamics of the constituents of macroscopic bodies. Rather it requires an assumption: the ergodic hypothesis. The approach adopted here can be traced back to the early works of Boltzmann [1], following an underground line (mostly unknown to the modern readers) that has received important later contributions by Helmholtz [2, 3], Boltzmann himself [4], Gibbs [5], P. Hertz [6], and Einstein [7]. I tried to present the material in an original and organic manner, casted in modern language and notation, and integrated with basic ergodic theory (as presented, e.g., in the textbook of Khinchin [8]), avoiding as much as possible unnecessary mathematical abstraction and complications.

The material presented is rather limited in comparison to what is customarily taught in a graduate course of statistical mechanics. It is meant only to cover the foundations of statistical ensembles and can be used as the first set of lectures of a complete course. It constitutes an ideal continuation of a graduate course in classical mechanics. Knowledge of basic calculus in many dimensions (including differential forms), thermodynamics, basic probability theory, and Hamiltonian mechanics are prerequisites for these lectures. Special attention is devoted to the Massieu potentials (the Legendre transforms of the entropy) which are most natural in statistical mechanics and also allow for a more direct treatment of the topic of ensemble equivalence. A number of exercises are scattered through the text, to stimulate learning and understanding. Many derivations, due to their simplicity, are skipped and proposed as exercises.

An important remark is in order for the teacher who wishes to use these lectures in a course of statistical mechanics. The theory presented here leads univocally to identify the thermodynamic entropy of an isolated system as the logarithm of the Liouville measure of the region of phase space enclosed by the energy hyper-surface, which is often referred to as volume entropy. That is in contrast with the standard textbook approach that postulates entropy as the logarithm of the invariant measure of said hyper-surface (the surface entropy), or, for the discrete system, the logarithm of the number of microstates according to Boltzmann's famous formula $S = k\ln W$. The latter formula is customarily used to treat various classic topics in the statistical mechanics of discrete systems (for example, when discussing the absence of phase transitions in the 1D Ising model, i.e., the argument of Landau and Peierls) which poses the question of how they should be treated within the present approach. In my experience, the formula $S = k\ln W$ could always be avoided, and that typically helped in improving the clarity. For example, the Landau and Peierls argument can be consistently presented as an application of the variational principle according to which the functional $f[\rho] = \mathrm{Tr}H\rho + T\mathrm{Tr}\rho\ln\rho$ is always larger or equal than the thermodynamic free energy F that is attained by the equilibrium canonical distribution $\rho \propto e^{-\beta H}$. The argument goes smoothly in this way and avoids the questionable habit, that permeates the whole literature, of calling non-equilibrium quantities, such as f or $-\mathrm{Tr}\rho\ln\rho$, with the name of thermodynamic potentials (free energy and entropy, respectively), which instead are, by definition, equilibrium quantities.

Chapter 1 is a quick review of the central statements of thermodynamics; the Massieu and standard potentials are introduced here. Chapter 2 presents the most minimal mechanical model of a thermodynamic system, namely a single particle in a 1D box. Thanks to the Helmholtz theorem that helps in realising that the thermo-dynamic structure is implicit in Hamiltonian mechanics. The chapter serves as an introduction to the ergodic theory, the microcanonical ensemble, and the generalised Helmholtz theorem, which are the subject of Chap. 3. Chapter 4 is devoted to the canonical ensemble, which is presented as the statistics that describes a system in weak contact with a large ideal gas. Chapter 5 deals with releasing the constraint on volume and presents the TP-ensemble as the ensemble that emerges from the canonical ensemble when the position of a movable piston is raised to the rank of a dynamical variable. Chapter 6 deals with the release of the constraint on the number of particles, namely, the grand-canonical ensemble. At variance with the standard textbook presentation, it is presented as an almost direct consequence of an assump-tion of "complete randomness" of the statistics of particle number (i.e., its Poisso-nian character is assumed), which is often implicit in standard derivations. Finally, Chap. 7 discusses the topic of ensemble equivalence, following a personal approach influenced by the works of Ruffo and others [9]. For all ensembles presented, their statistical-dynamical origin is highlighted, and thermodynamics is, according to the early Boltzmann spirit, never used as an input for their construction, rather it is presented as an aftermath.

The material is based on lectures delivered at the University of Florence in the academic years 2019/2020 and 2020/2021 as part of the course of Statistical Mechanics.

Pisa, Italy   Michele Campisi
July 2021

References

1. Boltzmann, L.: Wiener Berichte. **53**, 195 (1866)
2. von Helmholtz, H.: Journal für die reine und angewandte Mathematik. **97**, 111 (31 Dec. 1884). https://doi.org/10.1515/9783112342169-008
3. von Helmholtz, H.: Journal für die reine und angewandte Mathematik. **97**, 317 (31 Dec. 1884). https://doi.org/10.1515/9783112342169-019
4. Boltzmann, L.: Journal für die reine und angewandte Mathematik. **98**, 68 (1885). https://doi.org/10.1515/crll.1885.98.68
5. Gibbs, J.: Elementary Principles in Statistical Mechanics. Yale University Press, New Haven (1902)
6. Hertz, P.: Ann. Phys. (Leipzig) **338**, 225 (1910). https://doi.org/10.1002/andp.19103381202
7. Einstein, A.: Ann. Phys. (Leipzig) **339**, 175 (1911). https://doi.org/10.1002/andp.19113390111
8. Khinchin, A.: Mathematical Foundations of Statistical Mechanics, Dover, New York, (1949)
9. Campa, A., Dauxois T., Ruffo S.: Phys. Rep. **480**(3–6), 57 (2009). http://dx.doi.org/10.1016/j.physrep.2009.07.001

Contents

Chapter 1
The Constitutive Statements
of Thermodynamics

1.1 First and Second Law

1.1.1 First Law of Thermodynamics

In its differential form the first law of thermodynamics reads [1]:

$$dE = \delta Q + \delta W, \tag{1.1}$$

where dE is the change in internal energy, δQ is the heat added to the system and δW is the work done on the system during an infinitesimal transformation. The first law is the energy conservation law applied to a system in which there is an exchange of energy by both work and heat.

Of crucial importance for the understanding of the first law is that δQ and δW are, generally, *inexact* differentials, whereas dE is *exact*. The internal energy E is a *state variable*, namely a quantity that characterizes the thermodynamic equilibrium *state* of the system. On the other hand, W and Q are quantities that characterize thermodynamic energy *transfers* only and are not properties of the state of the system.

1.1.2 Second Law of Thermodynamics

Experience tells that:

Proposition 1.1 (The Heat Theorem) *For a quasi-static infinitesimal transformation, the differential $\delta Q / T$ is exact.*

where T is the temperature at which the heat δQ is exchanged. Proposition 1.1, also known as the *heat theorem* [2], is perhaps the most central statement of thermo-

© The Author(s), under exclusive license to Springer Nature Switzerland AG 2021
M. Campisi, *Lectures on the Mechanical Foundations of Thermodynamics*,
SpringerBriefs in Physics,
https://doi.org/10.1007/978-3-030-87163-5_1

dynamics: Although δQ is not an exact differential, it admits an integrating factor, namely the the inverse temperature.

This is equivalent to stating that that there exist a state function S, such that

$$\frac{\delta Q}{T} = dS. \tag{1.2}$$

The function S is called the *thermodynamic entropy* of the system.

In the following we adopt the convention of expressing temperature in units of energy. Accordingly, entropy is adimensional. Standard units can be covered by the change $T \rightarrow T/k_B$, $S \rightarrow k_B S$, with k_B Boltzmann's constant.

Proposition 1.1 can be expressed in an equivalent way also in *integral form*, by stating that the integral of $\delta Q/T$ along a reversible path connecting a state A to a state B in the state variables' space, does not depend on the path but only on its endpoints A and B. This in turn says that there exists a state function S (i.e. the thermodynamic entropy), such that

$$\int_A^B \frac{\delta Q}{T} = S(B) - S(A). \tag{1.3}$$

From Eq. (1.1) it is $\delta Q = dE - \delta W$. In general, the work is performed by changing a certain number of external parameters λ_i, e.g. volume, magnetic field, electric field, etc. . Then the work δW is given by $- \sum_i \Lambda_i d\lambda_i$, where Λ_i denote the corresponding conjugate forces, i.e., pressure, magnetization, electric polarization, respectively, etc. . Therefore it is:

$$\delta Q = dE + \sum_i \Lambda_i d\lambda_i. \tag{1.4}$$

In this case Proposition 1.1 can be re-expressed as:

$$\frac{dE + \sum_i \Lambda_i d\lambda_i}{T} = \text{exact differential} = dS, \tag{1.5}$$

or, in equivalent terms: *there exists a function $S(E, \lambda_1, \lambda_2, \dots)$ such that:*

$$\frac{\partial S}{\partial E} = \frac{1}{T}, \tag{1.6}$$

$$\frac{\partial S}{\partial \lambda_i} = \frac{\Lambda_i}{T}. \tag{1.7}$$

To use the language of vector fields, this says that the vector field $\boldsymbol{\Theta}(\mathbf{X})$, acting on the space

$$\mathbf{X} = (X_0, X_1, \dots, X_K) = (E, \lambda_1, \dots, \lambda_K) \tag{1.8}$$

with components

$$\Theta := (\Theta_0, \Theta_1, \ldots, \Theta_K) = (1, \Lambda_1, \ldots, \Lambda_K)/T \qquad (1.9)$$

is a *conservative* vector field. This is a highly non trivial fact. It implies the very existence of a *potential* function $S(\mathbf{X})$ such that

$$\Theta = \nabla S, \qquad (1.10)$$

where ∇ is the gradient operator in the space $\mathbf{X}$. In analogy with electric field and electric potential, we shall refer to Θ as the *thermodynamic field*, and to entropy S as the *thermodynamic potential*.

It is worth emphasizing that any inexact differential, like for example $\delta Q = dE + \sum_i \Lambda_i d\lambda_i$, does not enjoy the same property: it is generally not possible to find a function of state $g(\mathbf{X})$ such that $\partial g/\partial E = 1$ and $\partial g/\partial \lambda_i = \Lambda_i$.

Another remark is in order. While given a differential form defined on a 2-dimensional space it is always possible to find an integrating factor; that is not guaranteed in higher dimensional spaces. In other words, the very fact that the integrating factor $1/T$, and hence the potential S exist is highly non trivial. But that is what experience told Carnot, Clausius and Co.

Since for a system that is thermally insulated there is no heat exchange, an immediate consequence of Proposition 1.1 is that

Proposition 1.2 *For a quasi-static process occurring in a thermally isolated system, the entropy change between two equilibrium states is zero,*

$$\Delta S = 0. \qquad (1.11)$$

Proposition 1.1 refers to a quasi static transformation. The second law of thermodynamics is a statement regarding any thermodynamic transformation:

Proposition 1.3 (Second law) *For a generic process occurring in a system in thermal contact with a body of temperature T it is*

$$\delta Q/T \leq dS, \qquad (1.12)$$

with the equality holding when the transformation is quasi-static.

The latter proposition implies that

Proposition 1.4 *For a generic process occurring in a thermally insulated system it is*

$$0 \leq \Delta S. \qquad (1.13)$$

A crucial point that must not be overlooked is that Propositions 1.2 and 1.4 pertain to *thermally isolated* systems. This means that the system is not in contact with a thermal bath, by means of which one could in principle control either its temperature or its energy. Thus the processes mentioned therein are processes in which only the external parameters λ_i are varied in a controlled way and there is no control over the variable E. In Statement 1.2 the change of the parameter λ_i is so slow that at any instant of time the system is almost at equilibrium (quasi-static process). In Statement 1.4, this requirement is relaxed. Note also that in Statement 1.3 it is implied that that the infinitesimal transformation is between two equilibrium states.

Exercise 1.1 Consider the case of a single parameter λ with conjugate force Λ. Consider a closed path $\mathcal{C}$ in the space $\mathbf{X} = (E, \lambda)$. Show that the work W can be expressed as the flux Φ through the surface $\mathcal{S}$ bounded by the path of the "magnetic like" field $\mathcal{B}$

$$W = \Phi(\mathcal{B}) = \oint_{\mathcal{S}} d^2\,\mathbf{X}\mathcal{B}(\mathbf{X}), \tag{1.14}$$

where

$$\mathcal{B} = -\frac{\partial}{\partial E}\frac{\partial S/\partial \lambda}{\partial S/\partial E}. \tag{1.15}$$

Note that because of the first law $W = -Q$ for a closed path.   ■

1.2   Massieu Potentials

We restrict to the case of a single external parameter $\lambda_1 = V$ (volume) and conjugated force $\Lambda_1 = P$ (pressure), and we introduce the notation[1]:

$$\beta = 1/T, \tag{1.16}$$
$$\pi = P/T = \beta P. \tag{1.17}$$

The entropy $S(E, V)$ is a function of energy of the system and external parameter V and its differential reads:

$$dS = \beta dE + \pi dV. \tag{1.18}$$

This way of writing emphasises the existence of thermodynamically conjugated entropic pairs[2]

---

[1] The discussion can be repeated or generalised including the magnetic field H and magnetisation M, electric field E and polarisation $\mathcal{P}$, etc....

[2] The product of two variables that form an entropic pair has the dimension of entropy, i.e., with our convention of units, it is adimensional.

$$\beta \leftrightarrow E,$$
$$\pi \leftrightarrow V.$$

By Legendre transforming the entropy, one obtains new thermodynamic potentials, the Massieu potentials, where one or more among the variables E, V are exchanged with their conjugate.

In ordinary stable matter, the entropy is a concave function of (E, V),[3] implying:

$$\frac{\partial^2 S}{\partial E^2} < 0, \quad \frac{\partial^2 S}{\partial V^2} < 0, \quad \frac{\partial^2 S}{\partial E^2}\frac{\partial^2 S}{\partial V^2} - \left(\frac{\partial^2 S}{\partial E \partial V}\right)^2 < 0. \quad (1.19)$$

The Legendre transform are obtained accordingly through a maximisation.

1.2.1 The β, V Potential Ψ_F (Free Entropy)

The "free entropy", $\Psi_F(\beta, V)$, is the Legendre transform of entropy with respect to the change $E \leftrightarrow \beta$

$$\Psi_F(\beta, V) = \sup_E [S(E, V) - \beta E]. \quad (1.20)$$

Under the condition of stability $\partial^2 S/\partial E^2 < 0$ the Legendre transform then amounts to

$$\Psi_F(\beta, V) = S(E(\beta, V), V) - \beta E(\beta, V), \quad (1.21)$$

where $E(\beta, V)$ is the unique solution of the equation

$$\beta = \frac{\partial S}{\partial E}(E, V). \quad (1.22)$$

Unicity being a consequence of $\partial^2 S/\partial E^2 < 0$. The differential of the free entropy then reads

$$d\Psi_F = -Ed\beta + \pi dV. \quad (1.23)$$

Noting that $\partial^2 S/\partial E^2 = \partial T^{-1}/\partial E = -T^{-2}\partial T/\partial E = -1/(T^2 C_V)$, (where $C_V = (\partial T/\partial E)^{-1}$ is the heat capacity at constant V), we see that the condition $\partial^2 S/\partial E^2 < 0$, is equivalent to the condition:

$$C_V > 0. \quad (1.24)$$

---

[3] This is not always so, and we shall come back to this when discussing ensemble equivalence. A lot should be said regarding the topic of stability of matter. A good account is in the textbook of Callen [3].

1.2.2   The E, π Potential Ψ_H

Similarly, the Legendre transform of entropy with respect to V gives the Massieu potential

$$\Psi_H(E, \pi) = \sup_{V}[S(E, V) - \pi V]. \tag{1.25}$$

Under the concavity condition

$$\frac{\partial^2 S}{\partial V^2} < 0, \tag{1.26}$$

one can now define the function $V(E, \pi)$ as unique solution of

$$\pi = \frac{\partial S}{\partial V}(E, V), \tag{1.27}$$

and the Massieu potential gets the form

$$\Psi_H(E, \pi) = S(E, V(E, \pi)) - \pi V(E, \pi). \tag{1.28}$$

with differential

$$d\Psi_H = \beta dE - V d\pi. \tag{1.29}$$

1.2.3   The β, π Potential Ψ_G

By double Legendre transform one defines the potential

$$\Psi_G(\beta, \pi) = \sup_{E}[\Psi_H(E, \pi) - \beta E]. \tag{1.30}$$

Under the condition

$$\frac{\partial^2 \Psi_H}{\partial E^2} < 0, \tag{1.31}$$

it amounts to

$$\Psi_G(\beta, \pi) = \Psi_H(E(\beta, \pi), \pi) - \beta E(\beta, \pi), \tag{1.32}$$

where $E(\beta, \pi)$ is the unique solution of

$$\beta = \frac{\partial \Psi_H}{\partial E}(E, \pi). \tag{1.33}$$

The differential then reads

$$d\Psi_G = -Ed\beta - Vd\pi. \tag{1.34}$$

1.2.4 The β, V, α Massieu Grand Potential

Lastly, we introduce the Massieu grand potential. The idea is to explicitly express the N dependence of the potential $\Psi_F(\beta, V, N)$, and make a change of variables to the entropic partner of N. Let's call that $-\alpha$. Its differential reads

$$d\Psi_F = -Ed\beta + \pi dV - \alpha dN. \tag{1.35}$$

Typically $\Psi_F(\beta, V, N)$ is a concave function of N, accordingly the grand Massieu potential is defined as

$$\Psi_\Omega(\beta, V, \alpha) = \sup_N [\Psi_F(\beta, V, N) + \alpha N]. \tag{1.36}$$

Strict concavity of $\Psi_F(\beta, V, N)$ with respect to N, implies there exist a unique solution to:

$$-\alpha = \frac{\partial \Psi_F}{\partial N}. \tag{1.37}$$

Then the grand Massieu potential gets the form

$$\Psi_\Omega(\beta, V, \alpha) = \Psi_F(\beta, V, N(\beta, V, \alpha)) + \alpha N(\beta, V, \alpha). \tag{1.38}$$

The quantity α is related to the chemical potential by the relation:

$$\alpha = \frac{\mu}{T} = \beta\mu. \tag{1.39}$$

1.3 Common Thermodynamic Potentials

Massieu potentials, the Legendre transforms of the entropy, are rather uncommon in the literature, but, as we shall see, they emerge most naturally in the theory of statistical ensembles. Furthermore, the entropy S is the primordial potential. Accordingly its Legendre transforms are the fundamental potentials. For historical and practical reasons, much more common are the potentials obtained by Legendre transform of the internal energy.

1.3.1   The S, V Internal Energy E

Assuming $\partial S/\partial E = 1/T$ is non-negative, one first defines the function $E(S, V)$ by inverting $S(E, V)$ with respect to E and then proceeds to evaluate the Legendre transforms. By Eq. (1.5) it is:

$$dE = T dS - P dV, \tag{1.40}$$

which defines the set of thermodynamically conjugated energetic pairs[4]

$$T \leftrightarrow S,$$
$$P \leftrightarrow V.$$

In standard stable matter, the internal energy is a convex function of (S, V), implying:

$$\frac{\partial^2 E}{\partial S^2} > 0, \quad \frac{\partial^2 E}{\partial V^2} > 0, \quad \frac{\partial^2 E}{\partial S^2}\frac{\partial^2 E}{\partial V^2} - \left(\frac{\partial^2 E}{\partial V \partial S}\right)^2 > 0. \tag{1.41}$$

Accordingly now the Legendre transforms are implemented as infima, rather than suprema.

1.3.2   The T, V Helmholtz Free Energy F

The Helmholtz Free energy is defined as

$$F(T, V) = \inf_S [E(S, V) - TS]. \tag{1.42}$$

Under the assumption $\partial^2 E/\partial S^2 > 0$, it amounts to

$$F(T, V) = E(S(T, V), V) - T S(T, V), \tag{1.43}$$

with $S(T, V)$ being defined as the unique solution of

$$T = \frac{\partial E}{\partial S}(S, V). \tag{1.44}$$

The differential reads:

$$dF = -S dT - P dV. \tag{1.45}$$

---

[4] The product of two variables forming an energetic pair has the dimension of energy.

The concavity condition $\partial^2 E/\partial S^2 > 0$, is equivalent to the requirement that $C_V > 0$. Note that

$$F(T, V) = -T\Psi_F(1/T, V). \tag{1.46}$$

1.3.3 The S, P Enthalpy H

Similarly, the change $V \leftrightarrow P$ is achieved by the transform

$$H(S, P) = \inf_V[E(S, V) + PV], \tag{1.47}$$

Under the condition $\partial^2 E/\partial V^2 > 0$, it simplifies into

$$H(S, P) = E(S, V(S, P)) + PV(S, P) \tag{1.48}$$

with $V(S, P)$ the unique solution of

$$-P = \frac{\partial E}{\partial V}(S, V). \tag{1.49}$$

The differential reads

$$dH = TdS + VdP. \tag{1.50}$$

Noting that $V > 0$ by definition and that $\partial E/\partial V = -P$, the meaning of the stability condition is $-(\partial P/\partial V)_S > 0$, or $\chi_S > 0$, where $\chi_S = -(1/V)(\partial P/\partial V)_S$ is the adiabatic compressibility. It means that as the volume decreases adiabatically, the pressure increases (and vice-versa).

1.3.4 The T, P Gibbs Free Energy G

The Gibbs free energy is obtained by Legendre transform of the Enthalpy

$$G(T, P) = \inf_S[H(S, P) - TS]. \tag{1.51}$$

The condition of stability reads $\partial^2 H/\partial S^2 > 0$, leading to

$$G(T, P) = H(S(T, P), P) - TS(T, P), \tag{1.52}$$

where $S(T, P)$ is the unique solution of

$$T = \frac{\partial H}{\partial S}(S, P). \tag{1.53}$$

The differential reads, accordingly:

$$dG = -SdT + VdP. \tag{1.54}$$

The stability condition $\partial^2 H/\partial S^2$ implies $C_P > 0$. To see that, note that, at constant P, $TdS = \delta Q = C_P dT$. Therefore $\partial^2 H/\partial S^2 = \partial T/\partial S = T/C_P$, which under the assumption that $T > 0$, amounts to $C_P > 0$.

Note that

$$G(T, P) = -T\Psi_G(1/T, P/T). \tag{1.55}$$

1.3.5 The T, V, μ, Grand Potential

Consider the free energy F as a function of (T, V, N), and Legendre-transform with respect to N introducing its energetic partner μ to obtain the grand potential:

$$\Omega(T, V, \mu) = \inf_N[F(T, V, N) - \mu N]. \tag{1.56}$$

Assuming that F is strictly convex function of N, implies that the equation

$$\mu = \frac{\partial F}{\partial N}, \tag{1.57}$$

admits a unique solution $N(T, V, \mu)$ for fixed T, V, leading to:

$$\Omega(T, V, \mu) = F(T, V, N(T, V, \mu)) - \mu N(T, V, \mu). \tag{1.58}$$

For standard thermodynamical systems the potentials are extensive. In particular $G(T, P, N) = Ng(T, P)$. Since $\mu = \frac{\partial G}{\partial N}$, it is $\mu = g(T, P)$, hence $G = N\mu$, and $\Omega = F - N\mu$. Accordingly we get

$$\Omega = -PV. \tag{1.59}$$

Note that

$$\Omega(T, V, \mu) = -T\Psi_\Omega(1/T, V, \mu/T). \tag{1.60}$$

References

1. Fermi, E.: Thermodynamics. Dover, New York (1956)
2. Gallavotti, G.: Statistical Mechanics: A Short Treatise. Springer, Berlin (1999)
3. Callen, H.B.: Thermodynamics: An Introduction to the Physical Theories of Equilibrium Thermostatics and Irreversible Thermodynamics. Wiley, New York (1960)

Chapter 2
Minimal Mechanical Model of Thermodynamics

2.1 A Particle in a 1D Box

It is rather clear that the first law of thermodynamics can be understood as the macroscopic manifestation of the energy conservation law, that already holds at the microscopic level.

Is the "conservative" character of the thermodynamic field as well a consequence of the very Hamiltonian nature of the laws that govern the motion of the microscopic constituents of matter? The answer is "yes". Or better: "yes, provided one makes one further hypothesis, namely the *ergodic hypothesis*". We shall come back to this later.

In the meanwhile let us gain insight into the above question by considering the simplest model of a thermodynamic system one can think of: a particle in a 1D box.

We shall consider as external parameter the length L of the box, so that the Hamiltonian that describes the dynamics of the particle is:

$$H(x, p; L) = \frac{p^2}{2m} + U_{\text{box}}(x; L), \tag{2.1}$$

where

$$U_{\text{box}}(x; L) = \begin{cases} 0 & x \in (0, L) \\ \infty & \text{otherwise.} \end{cases} \tag{2.2}$$

The dynamics is very simple: The particle moves forth and back inside the box with constant speed $|v|$.

Let us now calculate the average force that the particle exerts on one wall in a period. The momentum transferred to a wall in a collision at fixed L is $\Delta p = 2m|v|$, hence the momentum transferred per unit time (let us call it P, like "pressure", the force conjugated to the "volume" L), is

© The Author(s), under exclusive license to Springer Nature Switzerland AG 2021
M. Campisi, *Lectures on the Mechanical Foundations of Thermodynamics*,
SpringerBriefs in Physics,
https://doi.org/10.1007/978-3-030-87163-5_2

$$P = \frac{2m|v|}{\tau} = \frac{m|v|^2}{L} = \frac{2E}{L} = P(E, L). \tag{2.3}$$

The heat differential, Eq. (1.4), reads, accordingly,

$$\delta Q = dE + \frac{2E}{L}dL. \tag{2.4}$$

First of all note that the above differential form, defined on the space (E, L) is not exact. To see that note that the differential form is not closed. Let us recall that (i) a differential form $\psi = M(x, y)dx + N(x, y)dy$ is said to be closed if $\partial_y M = \partial_x N$ and (ii) all exact forms are closed. Here the coefficient multiplying dE (i.e., the constant function 1) has null partial derivative with respect to L, while the coefficient multiplying dL, (i.e., the function $2E/L$), has non null partial derivative with respect to E. The differential form is not closed, hence not exact.

The question is then, does there exists an *integrating factor*, $g(E, L)$, such that

$$g(E, L)\delta Q = g(E, L)\left(dE + \frac{2E}{L}dL\right) \tag{2.5}$$

is exact? The answer is "yes". In fact it is always possible to find an integrating factor on 2-dimensional spaces, such as the present one [1]. As mentioned in Chap. 1 this is however not guaranteed to be always the case on larger spaces, and it will be important to keep this last point in mind. We also note that in fact there are infinitely many integrating factors, see Sect. 2.4.

Exercise 2.1 Before reading further, try and see if you can find an integrating factor $g(E, L)$ such that the differential form $g(E, L)dE + g(E, L)(2E/L)dL$ is exact. It might be helpful to first find an integrating factor g that makes the form closed, and then see if you can single out the associated primitive function $G(E, L)$ such that $dG = g(E, L)dE + g(E, L)(2E/L)dL$.   ∎

In order to find an integrating factor let the physics guide us. We have found an expression for "pressure" $P(E, L)$. Can we similarly write down a function $T(E, L)$ to which we can assign a meaning analogous to that of temperature? We know from basic kinetic theory of gases, that, in 3 dimensions, the temperature is 2/3 times the average kinetic energy of each single particle in the gas. Noting that in the present case the energy is all kinetic, and we are in 1 dimensions, we might assign to the following expression

$$T(E, L) = 2E, \tag{2.6}$$

the meaning of temperature. Is $1/T(E, L)$ an integrating factor? It is easy to show that the resulting differential form

$$\frac{1}{2E}dE + \frac{1}{L}dL \tag{2.7}$$

is closed. We recall that there exists a theorem according to which any closed differential form defined on an open simply connected set $\mathcal{A}$ is exact [1]. In the present case the differential form is defined onto the set $\mathcal{A} = \{(E, L) \in \mathbb{R}^2 | E > 0, L > 0\}$ which is open and simply connected, therefore the differential form is exact. Finding the primitive function $S(E, L)$ is straightforward: Since the differential is the sum of two differential forms defined each on $\mathbb{R}_+$, it suffices to integrate them separately and sum the results:

$$\frac{1}{2E}dE + \frac{1}{L}dL = d(\log\sqrt{E} + \log L + \text{const.}) = d(\log(\sqrt{E}L) + \text{const.}). \quad (2.8)$$

If we agree to name T the temperature and P the pressure, then we are obliged to agree to name the function

$$S(E, L) = \log\sqrt{E}L + \text{const.}, \quad (2.9)$$

the entropy of the system. What we have just proved is, in fact, that the gradient of $S(E, L)$ gives the thermodynamic field:

$$\frac{\partial S}{\partial E} = \frac{1}{T} = \frac{1}{2E}, \quad (2.10)$$

$$\frac{\partial S}{\partial L} = \frac{P}{T} = \frac{1}{L}. \quad (2.11)$$

2.1.1 Remarks

The equation $P = T/L$, or $PL = T$ is the ideal gas law, for the case of a single particle in 1D.

The condition of constant S defines the equation $E^{1/2}L = \text{const.}$, or using the equation of state, $PL^3 = \text{const}$ which is the adiabat's equation.

Exercise 2.2 Evaluate the specific heat at constant L, C_L, and the specific heat at constant P, C_P. Compare their ratio $\gamma = C_P/C_L$ with the exponent of the adiabat's equation. ∎

The entropy $S(E, L)$ has another crucial property, namely it is an *adiabatic invariant*.[1] To see that, let us imagine to vary L slowly in time according to a prescribed schedule $L(t)$. For simplicity let us assume a linear ramp $L(t) = L_0 + \dot{L}t$, with $\dot{L} > 0$ (expansion). Slowly means that the time scale of variation of L is long compared to the system's characteristic time. That is:

---

[1] In classical mechanics (and as well in quantum mechanics), a quantity that remains "unvaried" (what we mean by that will become more clear below) in slow processes is called an *adiabatic invariant* [2].

$$L/\dot{L} \gg \tau, \tag{2.12}$$

where $\tau = 2L/|v|$ is the period of the motion of the particle at fixed L, and v is the particle's velocity. The slowness condition then reads:

$$\dot{L}/|v| \ll 1. \tag{2.13}$$

Let us consider one collision. It is not difficult to show that if the particle velocity before the collision is v, then the post-collision velocity v' is $v' = -(v - 2\dot{L})$. Accordingly the speed $|v|$ decreases linearly with $\dot{L}$: $\Delta|v| = -2\dot{L}$.

Similarly, the energy changes by the amount

$$\Delta E = 2mv^2 \left[\left(\frac{\dot{L}}{v}\right)^2 - \frac{\dot{L}}{v} \right], \tag{2.14}$$

with the leading order being the linear one.

We now ask ourselves: does there exist a quantity which remains almost unvaried (that is whose change is at least quadratic in the slowness parameter $\dot{L}/v$) during the motion? For our particle in a box it is not difficult to see that the quantity $C = |v|L$ is an adiabatic invariant, in fact, in the time span of one period τ

$$\Delta C/\tau = |v'|(L/\tau + \dot{L}) - |v|L/\tau = (|v| - 2\dot{L})(L/\tau + \dot{L}) - |v|L/\tau = -2\dot{L}^2. \tag{2.15}$$

The quantity C admits a geometric interpretation: apart from a multiplicative factor it is the area A in phase space x, p enclosed by the trajectory that the particle draws for fixed L and velocity v: $A = 2|p|L = 2m|v|L$. Note that $|p|$ is a function of the energy $E = p^2/2m$ of the particle, hence A is a function of E and L:

$$A(E, L) = 2\sqrt{2mE}L. \tag{2.16}$$

Apart for a non important constant term, our entropy $S(E, L)$ is the log of the quantity $A(E, L)$. Since $A(E, L)$ is an adiabatic invariant, so is the entropy $S(E, L)$ as well. We thus see that the function $S(E, L)$ not only is the thermodynamic potential, but also is in perfect accordance with Proposition 1.2, according to which entropy is a quantity that remains unvaried in a slow process occurring in a thermal isolation, such as the one described here.

2.2   Helmholtz Theorem

The result in the previous section is a special case of a much more general result. Consider a generic 1D system with U-shaped potential $\phi(x; \lambda)$ that depends on an external parameter λ, see Fig. 2.1 for an example. Let

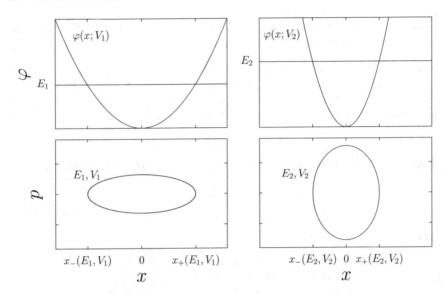

Fig. 2.1 Point particle in a harmonic potential $\varphi(x; V) = mV^2x^2/2$. Top left panel: potential energy at fixed $V = V_1$ as function of position x. Top right panel: potential energy at fixed $V = V_2 \neq V_1$ as function of position x. Bottom left panel: particle's orbit in phase space at energy E_1 and $V = V_1$. Bottom right panel: particle's orbit in phase space at energy E_2 and $V_1 = V_2$. The couple of parameters E, V, single out one closed orbit in the phase space, which represent an equilibrium state. Reproduced from Ref. [3], with the permission of the American Association of Physics Teachers

$$H(x, p; \lambda) = \frac{p^2}{2m} + \phi(x; \lambda) \tag{2.17}$$

be its Hamiltonian. Define:

$$T(E, \lambda) \doteq \left\langle \frac{p^2}{m} \right\rangle_t^{E,\lambda}, \tag{2.18}$$

$$\Lambda(E, \lambda) \doteq -\left\langle \frac{\partial H}{\partial \lambda} \right\rangle_t^{E,\lambda}, \tag{2.19}$$

where

$$\langle f \rangle_t^{E,\lambda} = \frac{1}{\tau} \int_0^\tau dt f(x(t), p(t)). \tag{2.20}$$

is the time average of the phase function $f(x, p)$ over the closed trajectory specified by E, λ. Here $\tau = \tau(E, \lambda)$ is the according period. For simplicity of notation we have dropped the explicit dependence of $\tau, x(t), p(t)$ on E, λ.

$\Lambda(E, \lambda)$ represents the average generalised force that the dynamical systems exerts on whatever external body keeps the parameter λ at its fixed value.

Exercise 2.3 Using the limit expression $U_{box}(x; L) = \lim_{\alpha \to \infty} \phi_\alpha$ where $\phi_\alpha = (x/L - 1/2)^\alpha$, show that $\Lambda_\alpha(E, L) = -\langle \partial_L \phi_\alpha \rangle_t^{E,L} \to 2E/L$. That is, show that the definition of Λ is consistent with that of average force (momentum transferred per unit time) exerted on the wall in the case a free particle in a 1D box. ∎

We are now ready to state the following[2]:

Theorem 2.1 (Helmholtz) *The differential* $(dE + \Lambda d\lambda)/T$ *is exact, that is,*

$$\frac{dE + \Lambda d\lambda}{T} = dS \tag{2.21}$$

and

$$S(E, \lambda) = \ln\left[2\int_{x_-(E,\lambda)}^{x_+(E,\lambda)} p(x; E, \lambda)dx\right]. \tag{2.22}$$

Here

$$p(x; E, \lambda) = \sqrt{2m(E - \phi(x; \lambda))} \tag{2.23}$$

denotes the momentum as a function of x, for the trajectory specified by E and λ, and $x_\pm(E, \lambda)$ are the according turning points.

Note that

$$A(E, \lambda) = 2\int_{x_-(E,\lambda)}^{x_+(E,\lambda)} p(x; E, \lambda)dx \tag{2.24}$$

is the area in phase space enclosed by the trajectory specified by E, λ.

Exercise 2.4 Prove Theorem 2.1. Hint: first prove that the period of motion is related to the area enclosed by the trajectory by the relation[3]

$$\tau(E, \lambda) = \frac{\partial A(E, \lambda)}{\partial E}. \tag{2.25}$$

∎

It is worth recalling that in general[4]:

---

[2] See Ref. [3].

[3] We shall come back to this later. This relation is discussed and derived in many textbooks of classical mechanics. For example Landau and Lifschitz classical mechanics textbook [2].

[4] See Landau and Lifschitz mechanics textbook [2] or any other textbook in classical mechanics.

Theorem 2.2 (Adiabatic Theorem) *Given a 1D system with U-shaped potential* $\phi(x; \lambda)$ *that depends on an external parameter* λ, *the specification of* E, λ *uniquely singles out a closed trajectory in phase space, whose enclosed area:*

$$A(E, \lambda) = \oint p \, dx, \tag{2.26}$$

is an adiabatic invariant.

2.3 First Encounter with Ergodicity and the Microcanonical Ensemble

Imagine we want to calculate the time average of a phase function $f(x, p)$ over the orbit specified by E and λ:

$$\langle f \rangle_t^{E,\lambda} := \frac{1}{\tau} \int_0^\tau dt f(x(t), p(t)). \tag{2.27}$$

Since $p = mv = m dx/dt$, the differential dt can be expressed as:

$$dt = m \frac{dx}{p(x)}, \tag{2.28}$$

where for simplicity we omitted the E, λ dependence of p. With this we obtain:

$$\langle f \rangle_t^{E,\lambda} = \frac{2m}{\tau} \int_{x_-}^{x_+} \frac{dx}{p(x)} f(x, p(x)), \tag{2.29}$$

where the factor 2 stems from the fact that the particle goes from x_- to x_+ in a half period, i.e., $\tau/2$. Now consider the following integral

$$\int dp \delta \left(p^2/2m + \varphi(x; \lambda) - E \right), \tag{2.30}$$

where δ denotes Dirac's delta function. Using the formula $\delta(f(p)) = \sum_i \delta(p - p_i)/|f'(p_i)|$, where the p_i's are the zeroes of $f(p)$, and $\int dp \delta(p - p_i) = 1$, we get:

$$\int dp \delta \left(p^2/2m + \varphi(x; \lambda) - E \right) = 2m/p(x). \tag{2.31}$$

Then Eq. (2.29) becomes:

$$\langle f \rangle_t = \frac{1}{\tau} \int dx \int dp \delta \left(p^2/2m + \varphi(x; \lambda) - E \right) f(p, x), \tag{2.32}$$

where the integration extremes $x_\pm$ need not be specified, being implied by the Dirac δ. The period τ is given by[5]:

$$\tau = \int_0^\tau dt = 2m \int_{x_-}^{x_+} \frac{dx}{p(x)}$$
$$= \int dx \int dp\delta \left(p^2/2m + \varphi(x;\lambda) - E \right). \tag{2.33}$$

Hence we arrive at

$$\langle f \rangle_t^{E,\lambda} = \langle f \rangle_\mu^{E,\lambda}, \tag{2.34}$$

where

$$\langle f \rangle_\mu^{E,\lambda} = \int dx \int dp\rho_\mu(x, p; E, \lambda)f(p, x) \tag{2.35}$$

denotes average over the following phase space distribution

$$\rho_\mu(x, p; E, \lambda) = \frac{1}{\tau(E, \lambda)}\delta \left(p^2/2m + \varphi(x;\lambda) - E \right). \tag{2.36}$$

From Eq. (2.36) it is clear that τ is the normalization of the distribution. The distribution $\rho_\mu(x, p; E, \lambda)$ is called the microcanonical distribution. Equation (2.34) says that the time average of a phase space quantity $f(x, p)$ over one period, is equal to its microcanonical average on phase space. This property is called ergodicity. We have just proved that all one-dimensional systems with a U-shaped potential are ergodic.

2.4 Non-uniqueness of Integrating Factors

We remark that, if there exist an integrating factor for a differential form, then, there exist infinitely many. Let $g(x, y)$ be an integrating factor for the form $M(x, y)dx + N(x, y)dy$, namely there exist a $G(x, y)$, such that

$$gMdx + gNdy = dG. \tag{2.37}$$

Consider now the function $H(x, y) = F(G(x, y))$, with F a differentiable function with non null derivative F'. Then

$$dH = F'(G)dG = F'(G)g(Mdx + Ndy), \tag{2.38}$$

---

[5] With this the solution of Exercise 2.4 is immediate.

which tells that $F'(G)g$ is an integrating factor. Due to the freedom in choosing F we see that there exist infinitely many integrating factors. The above argument can be repeated for differential forms in higher dimensions.

In light of the above, the Helmholtz theorem says that (i) the heat differential $dE + \Lambda(E, \lambda)d\lambda$ admits an infinite number of integrating factors each associated to a primitive function of the form $F(A(E, \lambda))$, which are all adiabatic invariants, (ii) among the integrating factors is the inverse of the average kinetic energy, which is associated to the primitive function $\ln A(E, \lambda)$.

2.5 Examples and Applications

2.5.1 Thermodynamics of a Harmonic Oscillator

Consider the following Hamiltonian of a one-dimensional harmonic oscillator with angular frequency λ

$$H(x, p; \lambda) = \frac{p^2}{2m} + \frac{m\lambda^2 x^2}{2}. \tag{2.39}$$

Exercise 2.5 (a) Calculate the area $A(E, \lambda)$ enclosed by the trajectory of energy E and angular frequency λ. Use Eq. (2.25) and check that the period of the orbit is, as expected, given by $\tau(E, \lambda) = 2\pi/\lambda$. (b) Use Theorem 2.1 to show that $T(E, \lambda) = E$ and $\Lambda(E, \lambda) = -E/\lambda$. (c) Show that the differential form $dE + \Lambda d\lambda$, with $\Lambda(E, \lambda)$ as in part (b) is not exact. (Hint: show that the differential form is not closed.). Show that the integral of $dE + \Lambda d\lambda$ over the rectangular path with corners $(E_0, \lambda_0), (E_0, \lambda_1), (E_1, \lambda_1), (E_1, \lambda_0)$, and $E_0 \neq E_1 \lambda_0 \neq \lambda_1$, is not zero. (d) Consider the differential form $\psi = (1/T)dE + (\Lambda/T)d\lambda$, with $\Lambda(E, \lambda)$ and $T(E, \lambda)$ as in part (b). Find a primitive function $S(E, \lambda)$ for ψ. Show that, apart from an additive constant, it is $S(E, \lambda) = \ln A(E, \lambda)$, as dictated by Theorem 1. Check that ψ is a closed differential form.   ■

2.5.2 The Peculiar Thermodynamics of a Log-Oscillator

Let us consider a system confined by a logarithmic potential, namely a so-called log-oscillator [4]:

$$H_{\log}(X, P) = \frac{P^2}{2M} + T \ln \frac{|X|}{b}, \tag{2.40}$$

where M is the mass, b is some positive constant with the dimension of length, and T is a positive constant with the dimension of energy. Figure 2.2 depicts some

Fig. 2.2 Black solid lines: Phase space trajectories of a log-oscillator at energies $E = 1/2, 1, 3/2, \ldots 9/2$, inner curves have lower energies. Red dashed line: the momentum distribution function, (2.45). Here $M = 1, T = 1$. Reprinted with permission from [5]. Copyright (2013) by the American Chemical Society

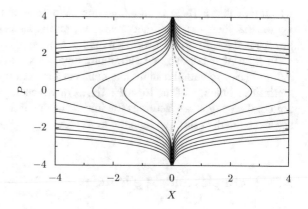

trajectories in phase space of different energies. Solving the equation $H_{\log}(X, P) = E$ for X, one sees that the trajectories are given by the equations:

$$X = \pm b\, e^{E/T} e^{-P^2/2MT} \, . \tag{2.41}$$

That is the trajectories have a Gaussian shape. Note that, accordingly, the maximal excursion grows exponentially with E/T: $X_{max} = b\, e^{E/T}$.

Exercise 2.6 (a) Show that the area $A(E)$ enclosed by the trajectory of energy E reads:

$$A(E) = 2b\sqrt{2\pi MT}\, e^{E/T} \, . \tag{2.42}$$

(b) Using the Helmholtz theorem show that

$$\langle P^2/M \rangle_E = (\partial S/\partial E)^{-1} = T. \tag{2.43}$$

∎

The latter equation expresses the major feature of the thermodynamics of a log-oscillator: all its trajectories have one and the same average kinetic energy, which is given by T, where T is the strength of the logarithmic potential. If we assign the average kinetic energy the meaning of temperature, this says that all trajectories, regardless of their energy have the same temperature. This fact is very peculiar: Consider for example the 1D harmonic oscillator, in this case $T(E) = E$ (see Exercise 2.5), namely the higher the energy, the higher the temperature. Similarly this is the case for a particle in a 1D box, where $T(E) = E/2$.

It therefore follows that the log-oscillator possesses a spectacular property: it has an *infinite* heat capacity; i.e.,

$$C(E) = (\partial T/\partial E)^{-1} = \infty, \tag{2.44}$$

thus, in a certain sense it is like a infinite thermal reservoir. We shall use this fact in Sect. 4.7.

Exercise 2.7 (a) Show that the probability density $f(P)$ to find the log-oscillator with momentum P, is given by the Maxwell distribution at temperature T:

$$f(P) = (2\pi MT)^{-1/2}e^{-P^2/(2MT)} \tag{2.45}$$

independent of its energy E. (b) Use this to prove once again that $T(E) = \langle P^2 \rangle_E / M = T$. ∎

The red dashed curve in Fig. 2.2 illustrates Eq. (2.45). When projecting the microcanonical distribution of the log-oscillator onto the P axis, the Maxwell distribution is obtained, regardless of the energy.

2.5.3 Particle in a Double Well Potential

Consider the following Hamiltonian describing a system in a symmetric double well

$$H(q, p) = \frac{p^2}{2m} - \frac{aq^2}{2} + \frac{bq^4}{4} + E_c \tag{2.46}$$

where the constant $E_c = a^2/(4b)$ is such that the minimum of energy is $E = 0$.

Exercise 2.8 (a) Prove that in the low energy limit $E \ll E_c$ it is $A(E) \propto E$. (b) Prove that in the high energy limit $E \gg E_c$ it is $A(E) \propto E^{3/4}$. ∎

Based on the results of the above exercise and the relation $1/T(E) = \partial_E \ln A(E)$, we have:

$$T(E) \simeq E, \quad E \ll E_c, \tag{2.47}$$

$$T(E) \simeq \frac{4}{3}E, \quad E \gg E_c. \tag{2.48}$$

Note that the line of constant energy $E = E_c$, in phase space is a separatrix. Namely it separates a region where the lines of constant energy are connected from a region where the lines of constant energy are disconnected. It is a known fact that the representative point takes an infinite time to traverse the trajectory at $E = E_c$, $\tau(E_c) = \infty$. Accordingly, the expectation value of the kinetic energy tends to zero for $E \to E_c$:

$$T(E) \to 0, \quad E \to E_c. \tag{2.49}$$

This means that the caloric curve, $T(E)$, is increasing for low E, reaches a maximum, becomes decreasing until it becomes zero at E_c, then becomes increasing again for $E > E_c$, and grows linearly for large E. The caloric curve $T(E)$ is sketched in Fig. 2.3.

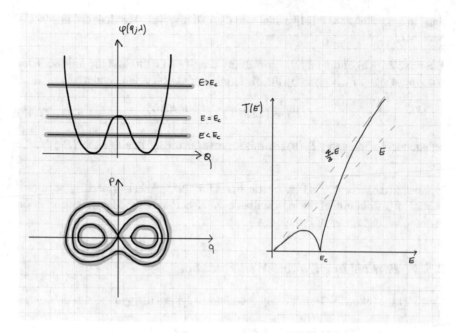

Fig. 2.3 Left panel: sketch of the double well potential in Eq. (2.46) (top) and trajectories in phase space at various energies (bottom). The orange trajectory, at $E = E_c$, is the separatrix. Right panel: Sketch of the caloric curve of a particle in the double well potential, Eq. (2.46)

Exercise 2.9 Use a software of your choice to plot the caloric curve. ∎

Note that at the local maximum it is $\partial T / \partial E = 0$, hence $C = \infty$: around the maximum, the system absorbs energy without changing much its temperature, like an infinite thermal bath or a log-oscillator.

Note that, below the critical energy E_c, there is a region where the caloric curve $T(E)$ is decreasing, namely where the heat capacity is negative $C(E) < 0$. In the microcanonical ensemble, that happens whenever there is a sufficiently fast increase of the phase space volume with increasing energy. Right below E_c, as the energy is increased, more and more of it goes into average potential energy, as the average kinetic energy diminishes: the energy you give is not used to speed up the motion, rather to slow it down, until you hit the separatrix. Above E_c the cooling mechanism does not hold any more and your system starts behaving in a standard way, namely by gaining speed as its energy increases further.

A negative microcanonical heat capacity is known to occur as well in self gravitating systems [6–8]. We shall come back to negative heat capacities in Chap. 7.

Note also that the caloric curve $T(E)$ is non-analytic at $E = E_C$, and so is the entropy $S(E)$, accordingly. Such non-analiticities are expected each time there is a sudden change of topology in the structure of the phase space [9]. We shall see more examples of this mechanism in Chap. 7.

Exercise 2.10 Consider a 3D particle in a $1/r$ potential. Show that $T(E) = -3E/2$, namely $C(E) = -3/2$ (This problem is discussed in Ref. [7]). ∎

References

1. Giusti, E.: Analisi matematica, vol. 2. Bollati Boringhieri (2003)
2. Landau, L.D., Lifshitz, E.M.: Mechanics. Addison-Wesley (1969)
3. Campisi, M., Kobe, D.H.: Am. J. Phys. **78**(6), 608 (2010). https://doi.org/10.1119/1.3298372
4. Campisi, M., Zhan, F., Talkner, P., Hänggi, P.: Phys. Rev. Lett. **108**, 250601 (2012). https://doi.org/10.1103/PhysRevLett.108.250601
5. Campisi, M., Hänggi, P.: J. Phys. Chem. B **117**(42), 12829 (2013). https://doi.org/10.1021/jp4020417
6. Lynden-Bell, D.: Physica A **263**(1–4), 293 (1999). https://doi.org/10.1016/S0378-4371(98)00518-4
7. Thirring, W.: Z. Phys. B **235**, 339 (1970). https://doi.org/10.1007/BF01403177
8. Campisi, M., Zhan, F., Hänggi, P.: EPL **99**, 60004 (2012). https://doi.org/10.1209/0295-5075/99/60004
9. Casetti, L., Kastner, M., Nerattini, R.: J. Stat. Mech. Theory Exp. **2009**(07), P07036 (2009). https://doi.org/10.1088/1742-5468/2009/07/p07036

Chapter 3
The Microcanonical Ensemble

3.1 Many-Body Model of Thermodynamics

We now can extend the previous treatment to systems of N particles. The Hamiltonian for an interacting system of N particles of masses m_i is

$$H(\mathbf{q}, \mathbf{p}; \lambda) = K(\mathbf{p}) + \varphi(\mathbf{q}; \lambda), \tag{3.1}$$

where $K(\mathbf{p}) = \sum_{i=1}^{s} p_i^2/2m_i$ is the kinetic energy, φ is the potential energy, and the coordinates $\mathbf{q} = \{q_i\}_{i=1}^{s}$ and their conjugate canonical momenta $\mathbf{p} = \{p_i\}_{i=1}^{s}$ are s-dimensional vectors. In 3 dimensions $s = 3N$.

In analogy with one-dimensional systems with a U-shaped potential, we define the microcanonical probability distribution as

Definition 3.1 (*microcanonical distribution*)

$$\rho_\mu(\mathbf{q}, \mathbf{p}; E, \lambda) = \frac{1}{\omega(E, \lambda)} \delta\left(E - H(\mathbf{q}, \mathbf{p}; \lambda)\right). \tag{3.2}$$

The quantity $\omega(E, \lambda)$ is the normalisation, also know as the structure function:

Definition 3.2 (*structure function*)

$$\omega(E, \lambda) = \int d\mathbf{q} d\mathbf{p} \delta\left(E - H(\mathbf{q}, \mathbf{p}; \lambda)\right). \tag{3.3}$$

Continuing the analogy with one-dimensional systems, we make the crucial assumption that for fixed E and λ, the phase space orbit followed by the system is uniquely determined. The intuitive meaning of this assumption is that the trajectory fills the hypersurface of constant energy:

© The Author(s), under exclusive license to Springer Nature Switzerland AG 2021
M. Campisi, *Lectures on the Mechanical Foundations of Thermodynamics*,
SpringerBriefs in Physics,
https://doi.org/10.1007/978-3-030-87163-5_3

$$\Sigma_{E,\lambda} = \{\mathbf{q}, \mathbf{p} \in \mathbb{R}^{2s} | H(\mathbf{q}, \mathbf{p}; \lambda) = E\}. \tag{3.4}$$

Namely we assume that there is only one trajectory with energy E for the given λ. This assumption allows us to use E and λ as state variables. In analogy with Eq. (2.34) we make the following:

Hypothesis 1 (*Ergodic hypotesis*) The time average $\langle f \rangle_t$ over the orbit specified by E and λ of any function $f(x, p)$ is equal to its microcanonical average $\langle f \rangle_\mu$, i.e.:

$$\langle f \rangle_t = \int d\mathbf{q} d\mathbf{p} \, \rho_\mu(\mathbf{q}, \mathbf{p}; E, \lambda) f(\mathbf{q}, \mathbf{p}) \doteq \langle f \rangle_\mu. \tag{3.5}$$

In analogy with the 1D case we define the temperature as:

$$T(E, \lambda) := \frac{2}{s} \langle K \rangle_t, \tag{3.6}$$

and the conjugate force as

$$\Lambda(E, \lambda) := -\left\langle \frac{\partial H}{\partial \lambda} \right\rangle_t. \tag{3.7}$$

We ask whether there exists a function $S(E, \lambda)$ such that

$$\frac{\partial S}{\partial E}(E, \lambda) = \frac{1}{T(E, \lambda)}, \quad \frac{\partial S}{\partial \lambda}(E, \lambda) = \frac{\Lambda(E, \lambda)}{T(E, \lambda)}. \tag{3.8}$$

The answer is given by the following theorem:

Theorem 3.1 (Helmholtz, Generalized) *A function $S(E, \lambda)$ satisfying Eq. (3.8) exists and is given by:*

$$S(E, \lambda) = \ln \Omega(E, \lambda), \tag{3.9}$$

where

$$\Omega(E, \lambda) := \int_{H(\mathbf{q},\mathbf{p}) \leq E} d\mathbf{q} d\mathbf{p} = \int d\mathbf{q} d\mathbf{p} \theta[E - H(\mathbf{q}, \mathbf{p})]. \tag{3.10}$$

The symbol $\theta(x)$ denotes the Heaviside step function. The proof is based on the

Theorem 3.2 (Microcanonical Equipartition Theorem)

$$\left\langle z_i \frac{\partial H}{\partial z_i} \right\rangle_\mu = \frac{\Omega(E, \lambda)}{\omega(E, \lambda)} \tag{3.11}$$

where z_i is any of the q's or the p's.

Let us first prove the Microcanonical Equipartition theorem:

$$
\begin{aligned}
\omega(E, \lambda) \left\langle z_i \frac{\partial H}{\partial z_i} \right\rangle_\mu &= \int d\mathbf{z}\, z_i \frac{\partial H}{\partial z_i} \delta[E - H(\mathbf{z}; \lambda)] \\
&= -\int d\mathbf{z}\, z_i \frac{\partial}{\partial z_i} \theta[E - H(\mathbf{z}; \lambda)] \\
&= -\int d\mathbf{z}^i \, (z_i \theta[E - H(\mathbf{z}; \lambda)])_{-\infty}^{+\infty} + \int d\mathbf{z} \theta[E - H(\mathbf{z}; \lambda)] \\
&= \Omega(E, \lambda),
\end{aligned}
\tag{3.12}
$$

where $\mathbf{z} = (\mathbf{q}, \mathbf{p})$, $\mathbf{z}^i = (z_0 \dots z_{i-1}, z_{i+1} \dots z_{2s})$, θ denotes Heviside step function, we have used $d\theta(x)/dx = \delta(x)$ to get to the second line, have integrated by parts in dz_i in the third line, and have used the fact that $H(\mathbf{z}; \lambda)$ grows without bounds as z_i grows (this being a consequence of the very form of H, with a quadratic kinetic part and a confining potential part). The microcanonical equipartition theorem follows immediately.

It says that each and all degrees of freedom share one and the same average value for the quantity $z_i \partial H/\partial z_i$. When z_i is one of the momenta p_j that quantity has the meaning of twice the kinetic energy associated with the jth degree of freedom. Since $K = \sum_i p_i^2/(2m_i)$, it is $\langle K \rangle_\mu = \frac{s}{2}\Omega/\omega$, hence, under the assumption of ergodicity:

$$
T(E, \lambda) = \frac{\Omega(E, \lambda)}{\omega(E, \lambda)}.
\tag{3.13}
$$

This is a crucial relation regarding ergodic Hamiltonian systems. It says that the temperature can be in fact expressed in geometrical terms: Ω is the volume of the portion of phase space enclosed by the hyper surface of constant energy Σ_E. Furthermore, from the relation $d\theta(x)/dx = \delta(x)$ it follows that

$$
\omega(E, \lambda) = \frac{\partial}{\partial E} \Omega(E, \lambda).
\tag{3.14}
$$

The proof of the Helmholtz theorem is straightforward now:

$$
\frac{\partial}{\partial E} \ln \Omega(E, \lambda) = \frac{1}{\Omega(E, \lambda)} \frac{\partial \Omega(E, \lambda)}{\partial E} = \frac{\omega(E, \lambda)}{\Omega(E, \lambda)} = \frac{1}{T(E, \lambda)},
\tag{3.15}
$$

which proves the first of Eq. (3.8) under the provision of the ergodic hypothesis. We further have

$$\frac{\partial}{\partial \lambda} \ln \Omega(E, \lambda) = \frac{1}{\Omega(E, \lambda)} \frac{\partial \Omega(E, \lambda)}{\partial \lambda}$$

$$= \frac{1}{\Omega(E, \lambda)} \int d\mathbf{z} \frac{\partial}{\partial \lambda} \theta[E - H(\mathbf{z}; \lambda)]$$

$$= -\frac{1}{\Omega(E, \lambda)} \int d\mathbf{z} \frac{\partial H}{\partial \lambda} \delta[E - H(\mathbf{z}\lambda)]$$

$$= -\frac{\omega(E, \lambda)}{\Omega(E, \lambda)} \int d\mathbf{z} \frac{\partial H}{\partial \lambda} \frac{\delta[E - H(\mathbf{z})]}{\omega(E, \lambda)}$$

$$= \frac{\omega(E, \lambda)}{\Omega(E, \lambda)} \left\langle -\frac{\partial H}{\partial \lambda} \right\rangle_\mu$$

$$= \frac{\Lambda(E, \lambda)}{T(E, \lambda)}, \tag{3.16}$$

where ergodicity has been used in the last equality. This proves the second of Eq. (3.8), thus completing the proof of Helmholtz' theorem.

The theorem can be immediately extended to include more parameters λ_i.

We draw the attention to the fact that, unlike temperature and conjugate force, the entropy $S(E, \lambda)$ is not written in the form of the time average of some phase function $f(\mathbf{q}, \mathbf{p})$.

A point that is important to stress is that the heat theorem, Eq. (1.2) which is the foundation of our construction, can fix the expression of microcanonical entropy, via the generalised Helmholtz theorem 3.1, only up to an additive constant $C(N)$, which can possibly depend on the number of particles N. This is because the derivatives of such a term with respect to the thermodynamic variables E, V would vanish and would accordingly not affect the thermodynamic field Θ. It also must be stressed that Ω is a quantity with the dimensions of $[\text{action}]^{3N}$, thus, for dimensional consistency it should be rescaled by some constant, that renders it a-dimensional, so that its logarithm, the entropy S, would be a-dimensional, in accordance with our choice of units, see Sect. 1.1.2. We shall ignore these issues for now, as they are not relevant for the subsequent discussion, and we shall come back to them in Chap. 6, when we will treat systems with varying number of particles.

Theorem 3.1 says that ergodic systems constitute ideal mechanical models of thermodynamics. One can define their state variables by E and λ as in thermodynamics. Moreover, one can define their temperature and conjugate force straightforwardly as functions of the state variables. Surprisingly, the heat differential $(dE + \Lambda d\lambda)/T$ is exact, allowing for a consistent mechanical expression of entropy S, independent of the size of the system: Not all macroscopic properties of large systems follow from their size being big!

This shows that the conservative nature of the thermodynamic field Θ is a consequence of ergodic Hamiltonian dynamics of the constituents of physical bodies, regardless of their size. Similarly the first law of thermodynamics follows from energy conservation that holds at the microscopic level.

3.2 Remarks on Ergodicity

In the previous chapter we have remarked that all 1-dimensional systems confined in a U-shaped potential are ergodic. The situation is extremely more complicated in phase spaces with dimension larger than 2. Ergodic theory is a whole branch of mathematical physics that studies the notion of ergodicity in many-body systems, and the conditions under which it holds. We are not entering into this fascinating field, but only restrict ourselves to make some observations and state some central results.

It must be remarked that ergodicity has been proven formally for a very small number of models. Its proof is generally a most formidable task.[1] This looks, at first sight, as a very unpleasant situation: On one hand ergodicity is the assumption that makes the connection between microscopic dynamics and the second law of thermodynamics (at least for the part that concerns the statement that the thermodynamic field is conservative) possible; while, on the other, we are able to prove ergodicity only for a handful of very specific cases. This appears as a rather weak link for our macro-to-micro reduction program.

It is worth noticing however that, in fact, for the proof of the generalised Helmholtz theorem to be completed, one needs that only certain phase functions, namely the kinetic energy K and the forces $-\partial H/\partial \lambda_i$ be ergodic, namely that Eq. (3.5) holds only for a very small set of functions, rather than for any function f. Hence clearly the ergodic hypothesis, as formulated above is too strict of a condition for the generalised Helmholtz theorem, which holds as well under much looser conditions. Furthermore, despite the embarrassing situation, it is pretty widely accepted, based on numerical evidence, that many systems of physical interest e.g., gas of particles with short range interactions and a repulsive core behave, for all practical proposes, as if they were ergodic. The problem is, after all, not as serious as one might have thought at first sight.

3.2.1 Metric Indecomposability

The following result of ergodic theory might help us gaining a better understanding of the ergodic hypothesis. Consider a subset $\Pi \subset \mathbb{R}^{2s}$ of the phase space.

Definition 3.3 $\Pi \subset \mathbb{R}^{2s}$ is said to be *metrically indecomposable* if:

- Π is invariant under the Hamiltonian flow
- Π cannot be written as $\Pi = \Pi_1 \cup \Pi_2$ where $\Pi_{1,2} \subset \mathbb{R}^{2s}$ are disjoint $\Pi_1 \cap \Pi_2 = \emptyset$, are themselves invariant under the Hamiltonian flow, and have non-zero measure.

---

[1] A nice and accessible account on the modern ergodic theory was given some years ago by Lebowitz and Penrose [1].

Here by measure we mean the Euclidean measure in $\mathbb{R}^{2s}$:

$$\mathcal{L}(\Pi) = \int_{z \in \Pi} d\mathbf{z}. \tag{3.17}$$

By virtue of Liouville theorem, $\mathcal{L}$ is an invariant measure, that is: $\mathcal{L}(\Pi) = \mathcal{L}(\Pi_t)$, where Π_t is the set Π evolved for a time t, according to Hamiltonian evolution. We shall refer to $\mathcal{L}$ as the Lioville measure.

Let us now consider the initial condition $\mathbf{z}_0$ and let $\mathbf{z}_t$ be the evolved at time t. Let $\hat{f}(\mathbf{z}_0)$ be the time average of $f(\mathbf{z})$ over the trajectory having $\mathbf{z}_0$ as initial condition. The following holds

Theorem 3.3 *If Π is metrically indecomposable, then*

$$\hat{f}(\mathbf{z}_0) =_{a.e.} \frac{\int_\Pi d\mathbf{z} f(\mathbf{z})}{\int_\Pi d\mathbf{z}}. \tag{3.18}$$

Here the subscript $a.e.$ means "almost everywhere". The interested reader may find a proof in the book of Khinchin [2].

A consequence of theorem 3.3 is that, if the surface of constant energy Σ_E is metrically indecomposable, then, at energy E, (apart from at most a zero measure set of initial conditions) the ergodic hypothesis, Eq. (3.5) holds.

To see that consider the set

$$V_E = \{\mathbf{z} \in \mathbb{R}^{2s} | H(\mathbf{z}) \le E\}, \tag{3.19}$$

where for simplicity we drop the label λ. Now consider the set

$$\Delta V = \{\mathbf{z} \in \mathbb{R}^{2s} | E \le H(\mathbf{z}) \le E + \Delta E\}, \tag{3.20}$$

that is a shell of energy between E and $E + \Delta E$. ΔV is invariant under Hamilonian flow but it is not metrically indecomposable because it can be decomposed into two positive measure invariant sets, e.g., the shell of energy in $[E, E + \Delta E/2]$ and the shell of energy in $]E + \Delta E/2, E + \Delta E]$. We take then the limit $\Delta E \to 0$ of ΔV and assume the limit set, namely Σ_E, is metrically indecomposable. We obtain

$$\hat{f}(\mathbf{z}_0) =_{a.e.} \frac{\lim_{\Delta E \to 0} \int_{\Delta V} d\mathbf{z} f(\mathbf{z})}{\lim_{\Delta E \to 0} \int_{\Delta V} d\mathbf{z}} = \frac{\frac{d}{dE} \int_{V_E} d\mathbf{z} f(\mathbf{z})}{\frac{d}{dE} \int_{V_E} d\mathbf{z}}. \tag{3.21}$$

Writing $\int_{V_E} d\mathbf{z} f(\mathbf{z}) = \int d\mathbf{z} \theta[E - H(\mathbf{z})] f(\mathbf{z})$, it is straightforward to see that:

$$\frac{d}{dE} \int_{V_E} d\mathbf{z} f(\mathbf{z}) = \int d\mathbf{z} f(\mathbf{z}) \delta[E - H(\mathbf{z})] \tag{3.22}$$

$$\frac{d}{dE} \int_{V_E} d\mathbf{z} = \int d\mathbf{z} \delta[E - H(\mathbf{z})] = \omega(E). \tag{3.23}$$

Thus, we get

$$\hat{f}(\mathbf{z}_0) =_{a.e.} \int d\mathbf{z} f(\mathbf{z}) \frac{\delta[E - H(\mathbf{z})]}{\omega(E)} = \langle f \rangle_\mu, \tag{3.24}$$

which proves our statement.

3.2.1.1 Remarks

One important point to notice is that in order for Σ_E to be metrically indecomposable the equations of motion must have *no first integrals besides the energy*. To see that, first of all note that, in general Σ_E is invariant under the Hamiltonian flow. Then consider an integral of motion $I(\mathbf{z})$ which is functionally independent from $H(\mathbf{z})$, see Fig. 3.1. Let us say that $I_1 < I_2$ are the smallest and largest values taken by I on Σ_E. Take some intermediate value $\bar{I} \in]I_1, I_2[$, and consider the sets $\Sigma_1 = \{\mathbf{z} \in \Sigma_E | I(\mathbf{z}) < \bar{I}\}$, $\Sigma_2 = \{\mathbf{z} \in \Sigma_E | I(\mathbf{z}) \geq \bar{I}\}$. The sets have non-zero measure, are disjoint, and $\Sigma_1 \cup \Sigma_2 = \Sigma_E$. Furthermore they are invariant under Hamilton flow. Hence Σ_E is not metrically indecomposable. Intuitively, the condition of ergodicity occurs when the trajectory densely fills the whole surface Σ_E. If there are further integrals of motion $I, J, \dots$ the trajectory explores only the subset where $I, J, \dots$ are constant, and is evidently not able to explore the whole surface.

Considering a gas of noninteracting particles in a cubic box, clearly there are as many integrals of motion as the number s of degrees of freedom, namely $I_i = p_i^2$. When that is the case we say the system is integrable. So, non-interacting system are the "least ergodic" systems one can think of. They are certainly not good models of thermodynamic behaviour. Interactions are essential for the emergence of the thermodynamic structure.

Another situation which may break metric indecomposability is when there is a "spontaneous symmetry breaking".[2] The simplest case where that happens is for a single 1-d particle in a double well potential, as in the quartic potential in Eq. (2.46):

$$H(q, p) = \frac{p^2}{2m} - \frac{aq^2}{2} + \frac{bq^4}{4} + E_c. \tag{3.25}$$

The Hamiltonian has the symmetry $q \leftrightarrow -q$, which is accordingly a symmetry of Σ_E. For $a > 0$ the potential energy has two symmetric minima located at $q = \pm\sqrt{a/b}$

---

[2] By this we mean that while the Hamiltonian has a certain symmetry, the solution of the equation of motion does not have it.

Fig. 3.1 Integrals of motion
and ergodicity

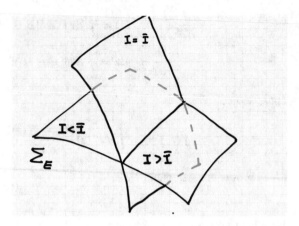

and a maximum at $q = 0$ corresponding to $E = E_c$, see Fig. 2.3. Below E_c, Σ_E
breaks in two disconnected closed curves in the phase space (q, p), call them Σ_E^L
and Σ_E^R, which are one the mirror image of the other (with respect to the change
$q \leftrightarrow -q$). If the system is initially on one of them, it can never explore the other one,
hence Σ_E is not metrically indecomposable. Consider the observable $f(q, p) = q$.
Then clearly, for energy below E_c, its time average is non null, and equal to $\pm c$, with
some constant $c > 0$, depending on whether the particle is in the right or left curve,
respectively. Those averages are equal to phase averages over the corresponding
metrically indecomposable component Σ_E^L and Σ_E^R, respectively. However, because
of symmetry, the average over Σ_E is null, hence the ergodic hypothesis, Eq. (3.5),
breaks down.

Note however that the ergodic hypothesis remains valid for those observables that
have the symmetry that is being broken. For example the kinetic energy $p^2/2m$
is ergodic. Similarly, considering a (or b) as an external parameter, its conju-
gated force $-\partial H/\partial a = q^2/2$, (or $-\partial H/\partial b = -q^4/4$) is ergodic. We thus see that,
despite ergodicity generally breaks, it remains valid for the observables entering the
Helmholtz theorem, which accordingly remains valid.

As we shall see, the critical energy where the breaking occurs is a point of nonanaliticity of $S(E)$ and hence signals a sort of microcanonical phase transition. We
will re-encounter such non-analiticities again in Chap. 7.

The above discussion should also clarify our statement that all one dimensional
systems with a U shaped potential are ergodic. If the potential is U-shaped there is
only one trajectory per energy specification, and the system explores the whole Σ_E.
Note the fact that such 1-dimensional systems are very special in that they are at the
same time integrable and ergodic.

3.2.2 The Invariant Measure

A crucial result is the following [2]

Theorem 3.4

$$\frac{d}{dE} \int_{V_E} d\mathbf{z} f(\mathbf{z}) = \int_{\Sigma_E} \frac{d\sigma}{|\nabla H|} f(\mathbf{z}) . \tag{3.26}$$

The symbol $d\sigma$ stands for the surface element on Σ_E, and ∇ denotes the gradient in phase space.

To prove that consider the hyper surface Σ_E. Now consider a surface element $d\sigma$ on Σ_E, see Fig. 3.2, left panel. Then a volume element $d\mathbf{z}$ is given by $d\sigma dh$ where dh is the differential of the local coordinate perpendicular to $d\sigma$. Now, since $d\sigma$ is on Σ_E, which is a surface of constant energy, the direction perpendicular to $d\sigma$ is given by the versor $\mathbf{n} = \nabla H/|\nabla H|$. Moving of dh in that direction amounts to increasing the energy by the quantity $dE = |\nabla H| dh$. In sum,

$$d\mathbf{z} = d\sigma dh = d\sigma dE/|\nabla H| , \tag{3.27}$$

then

$$\frac{d}{dE} \int_{V_E} d\mathbf{z} f(\mathbf{z}) = \frac{d}{dE} \int^E dE' \int_{\Sigma_{E'}} d\sigma f(\mathbf{z})/|\nabla H| = \int_{\Sigma_E} \frac{d\sigma}{|\nabla H|} f(\mathbf{z}). \tag{3.28}$$

The measure $d\sigma/|\nabla H|$ on the surface of constant energy is called *the invariant measure*. The reason is that given a set $A \subseteq \Sigma_E$ the measure:

$$M(A) = \int_A \frac{d\sigma}{|\nabla H|} \tag{3.29}$$

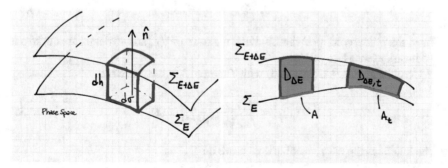

Fig. 3.2 Left panel: Volume element in phase space. Right panel: The invariant measure

is invariant under the Hamiltonian flow, namely, if A_t is the evolved of A after a time t, it is

$$\mathcal{M}(A) = \mathcal{M}(A_t). \tag{3.30}$$

To prove Eq. (3.30) consider some set $D \subset \mathbb{R}^{2s}$, and let us take a non-empty slice of it $D_{\Delta E} = D \cap \Delta V$, by choosing appropriate E and ΔE, see Fig. 3.2 (ΔV was defined in Eq. (3.20). Let $A = D_{\Delta E} \cap \Sigma_E$. We want to show that, since the $\mathcal{L}$ measure of $D_{\Delta E}$ is invariant, then the $\mathcal{M}$ measure of A is also invariant. Namely, we want to show that the Liouville measure $\mathcal{L}$ in phase space induces the measure $\mathcal{M}$ on the hyper surfaces Σ_E. Let us introduce the characteristic function $\chi(\mathbf{z})$ which takes on the value 1 if $\mathbf{z} \in D_{\Delta E}$, and takes on the value 0 otherwise. We have

$$\mathcal{L}(D_{\Delta E}) = \int_{\Delta V} d\mathbf{z} \chi(\mathbf{z}) = \int_{V_{E+\Delta E}} \chi \, d\mathbf{z} - \int_{V_E} \chi \, d\mathbf{z}, \tag{3.31}$$

which is invariant, no matter how small ΔE is. By taking the limit $\Delta E \to 0$, we obtain that

$$\frac{d}{dE} \int_{V_E} d\mathbf{z} \chi(\mathbf{z}) \tag{3.32}$$

is invariant. But, by virtue of Theorem 3.4 the quantity above is $\mathcal{M}(A)$ which proves Eq. (3.30).

3.2.2.1 Remarks

It is immediate to see that:

$$\mathcal{L}(V_E) = \Omega(E), \tag{3.33}$$

$$\mathcal{M}(\Sigma_E) = \omega(E). \tag{3.34}$$

Thus, the phase volume is the Liouville measure of V_E, and structure function ω is the invariant measure of Σ_E.

Combining Eqs. (3.22), (3.23) with Theorem 3.4, we can write

$$\langle f \rangle_\mu = \int_{\Sigma_E} \frac{d\sigma}{\omega(E)|\nabla H|} f(\mathbf{z}), \tag{3.35}$$

which defines the microcanonical measure on Σ_E

$$d\mu = \frac{d\sigma}{\omega(E)|\nabla H|}. \tag{3.36}$$

Note that, at variance with what is stated in most textbooks, the microcanonical measure is not a flat measure on the hyper surface of constant energy. Namely not all points on Σ_E have the same weight in the calculation of the microcanonical average. One often reads incorrect statements like "all microscopic states corresponding to the same macroscopic condition have same weight in the microcanonical ensemble" or similar. Equation (3.35) says the opposite: different microscopic states $\mathbf{z}$ having the same energy E, will be weighted differently, depending on the value of $|\nabla H|$. In other words the microcanonical ensemble is not homogeneously distributed on Σ_E. Where $|\nabla H|$ is higher, the ensemble density is lower and vice-versa. What is the physical meaning of this inhomogeneity? Note that, by Hamilton's equation $\nabla H = (\nabla_q H, \nabla_p H) = (-\dot{\mathbf{p}}, \dot{\mathbf{q}})$. So, $|\nabla H|$ is the speed $|\mathbf{v}|$ of the Hamilton flow at $\mathbf{z}$. Accordingly the faster the flow the less the density of the ensemble, see Fig. 3.3. Loosely speaking, the microcanonical measure of some set A, is a measure of the fraction of time that a representative point spends in it, during its run towards filling the whole hypersurface. We now better understand why microcanonical averages coincide with time averages.

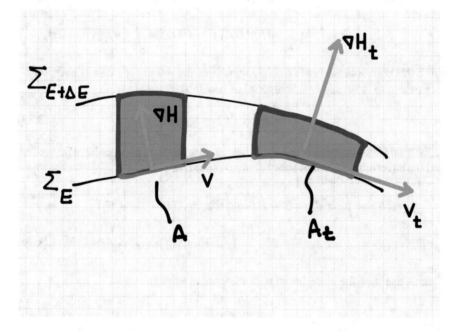

Fig. 3.3 The figure helps understanding the invariant measure. The blue volumes are equal by Liouville theorem. Their heights $h = \Delta E/|\nabla H|$, $h_t = \Delta E/|\nabla H_t|$ are inversely proportional to the speed $|\mathbf{v}| = |\nabla H|$, hence their bases, A, and A_t have areas (orange) $\int_A d\sigma$, $\int_{A_t} d\sigma$ that are proportional to $|\nabla H|$. Therefore, by renormalising them by $|\nabla H|$ the invariant measure $\int_A d\sigma/|\nabla H|$ is obtained. Imagine to have n representative points homogeneously distributed in A, and let them evolve for a time t. Their surface density would be smaller in A_t where the flow speed is higher

Exercise 3.1 For a 1d system in a U-sahped potential show that the microcanonical measure of a set A on the surface of constant energy Σ_E is the fraction of time that the system representative point spends in it, over a full period τ.   ∎

3.3   Properties of the Structure Function

Throughout these lectures, whenever not specified, it is assumed that the Hamiltonian is bounded from below and we set conventionally its minimum at value 0. Accordingly $\Omega(E)$ is defined on the positive axis, and $\Omega(E) \geq 0$. Furthermore, it is a strictly increasing function, namely $\Omega'(E) = \omega(E) > 0$. Accordingly, the function $T(E) = \Omega/\omega > 0$, is non-negative. This means that in the microcanonical ensemble, temperature is positive definite.

The structure function $\omega(E)$ is useful for the evaluation of phase integrals of phase functions of the form $f(\mathbf{z}) = g(H(\mathbf{z}))$:

$$\int d\mathbf{z} g(H(\mathbf{z})) = \int d\mathbf{z} \int_0^\infty dE g(H(\mathbf{z}))\delta(E - H(\mathbf{z})) \tag{3.37}$$

$$= \int d\mathbf{z} \int_0^\infty dE g(E)\delta(E - H(\mathbf{z})) = \int_0^\infty dE g(E)\omega(E).$$

From this equation, we see that $\omega(E)$ represents the *density of states* at energy E.

Imagine to have a system composed of two subsystems. Let

$$H(\mathbf{z}) = H_1(\mathbf{z}_1) + H_2(\mathbf{z}_2) \tag{3.38}$$

be the total Hamiltonian. Let ω, ω_i, be the structure function of the composite system, and the structure function of subsystem i, respectively. It is:

$$\omega(E) = \int d\mathbf{z}_1 d\mathbf{z}_2 \delta[E - H] = \int d\mathbf{z}_1 d\mathbf{z}_2 \delta[E - H_1 - H_2]$$

$$= \int d\mathbf{z}_1 \omega_2(E - H_1) = \int_0^E dE_1 \omega_1(E_1)\omega_2(E - E_1). \tag{3.39}$$

That is, denoting the convolution with the symbol $*$,

$$\omega = \omega_1 * \omega_2. \tag{3.40}$$

Similarly, it can be proved that

$$\Omega = \Omega_1 * \omega_2 = \omega_1 * \Omega_2. \tag{3.41}$$

Imagine the composite system is at energy E and distributed according to the microcanonical distribution:

$$\rho(\mathbf{z}; E) = \frac{\delta[E - H(\mathbf{z})]}{\omega(E)}. \tag{3.42}$$

We ask what is the probability density $p(\mathbf{z}_1; E)$ to find subsystem subsystem 1 at point $\mathbf{z}_1$. Using the rules of probability, that is calculated by integrating out $\mathbf{z}_2$:

$$p(\mathbf{z}_1; E) = \int d\mathbf{z}_2 \frac{\delta[E - H(\mathbf{z})]}{\omega(E)}. \tag{3.43}$$

It is straightforward to see that

$$p(\mathbf{z}_1; E) = \frac{\omega_2(E - H_1(\mathbf{z}_1))}{\omega(E)}. \tag{3.44}$$

This implies the following important fact: the functional shape of the distribution of a subsystem is dictated by the shape of the structure function of its complement.

Exercise 3.2 Prove Eq. (3.44). ∎

From Eq. (3.44) follows that the probability density function $P(E_1; E)$ of finding subsystem 1 at et energy E_1, given that the total system is at energy E, reads:

$$P(E_1; E) = \int d\mathbf{z}_1 p(\mathbf{z}_1; E) \delta[E_1 - H_1(\mathbf{z}_1)] = \frac{\omega_2(E - E_1)\omega_1(E_1)}{\omega(E)}. \tag{3.45}$$

3.4 Ideal Gas

Let us now come to the prime example. Namely the ideal gas. We consider N particles in a 3D box of volume V. The Hamiltonian reads:

$$H(\mathbf{q}, \mathbf{p}; V) = \sum_i^{3N} \frac{p_i^2}{2m} + U_{\text{box}}(\mathbf{q}; V), \tag{3.46}$$

where $U_{\text{box}}(\mathbf{q}; V)$ is null if all particles are in the volume V, and ∞ otherwise. As mentioned above, this is certainly not an ergodic system. Nonetheless we can evaluate the phase volume Ω and the quantity $S = \ln \Omega$. Let us pretend for a while it represents the entropy, and derive the according "thermodynamics". We shall later comment on whether that makes sense. We have:

$$\Omega(E, V) = \int d\mathbf{q} d\mathbf{p} \, \theta \left[E - \sum_i \frac{p_i^2}{2m} - U_{\text{box}}(\mathbf{q}; V) \right] \tag{3.47}$$

The configurational integral (i.e., the integration in $d\mathbf{q}$) is easily performed and amounts to V^N. Then

$$\Omega(E, V) = V^N \int d\mathbf{p} \, \theta \left[2mE - \sum_i p_i^2 \right],\tag{3.48}$$

where we used the property $\theta(ax) = \theta(x)$ for $a > 0$. The integral is the hyper volume $\mathcal{V}_n(R)$ of a sphere of radius $R = \sqrt{2mE}$ in dimension $n = 3N$. Using the formula:

$$\mathcal{V}_n(R) = \frac{\pi^{n/2} R^n}{\Gamma(1 + n/2)},\tag{3.49}$$

where Γ denotes the Gamma-function, we arrive at:

$$\Omega(E, V) = \frac{\pi^{3N/2}}{\Gamma(1 + 3N/2)} (2m)^{3N/2} V^N E^{3N/2}.\tag{3.50}$$

Taking the derivative with respect to E, we get:

$$\omega(E, V) = \frac{3N}{2} \frac{\pi^{3N/2}}{\Gamma(1 + 3N/2)} (2m)^{3N/2} V^N E^{3N/2-1} = \frac{3N\Omega(E, V)}{2E}.\tag{3.51}$$

Using Eqs. (3.8, 3.9) we obtain:

$$E = \frac{3N}{2} T,\tag{3.52}$$

$$PV = NT,\tag{3.53}$$

which are the equations of state of an ideal gas. It follows that the heat capacity at constant volume is given by

$$C_V = \left(\frac{\partial T}{\partial E} \right)^{-1} = \frac{3N}{2}.\tag{3.54}$$

3.4.1 Remarks

Notwithstanding the Hamiltonian is integrable, hence drastically not ergodic, we have nonetheless found the thermodynamics of the ideal gas. Why is it so? In a real gas, regardless of how dilute it is, the particles interact with each other, so the Hamiltonian in Eq. (3.46) is an idealisation. What we have learned is that, despite those interactions exist, and are crucial for ergodicity to be reached, they in fact can be neglected in the calculations, as long as we are concerned with the ideal gas thermodynamics. When studying real gases with equations of state that depart from ideal gas case we must certainly include the interactions into the calculations in order to find the deviations from the ideal gas law.

Exercise 3.3 (a) Given a generic Hamiltonian of the type

$$H(\mathbf{q}, \mathbf{p}) = \sum_i^{3N} \frac{p_i^2}{2m} + \phi(\mathbf{q}), \tag{3.55}$$

show that the phase volume can be written as the convolution,

$$\Omega = \Omega_c * \omega_k, \tag{3.56}$$

of the so-called configuration integral,

$$\Omega_c(E) = \int d\mathbf{q}\,\theta[E - \phi(\mathbf{q})], \tag{3.57}$$

and the kinetic structure function reading:

$$\omega_k(E) = \frac{3N}{2} \frac{\pi^{3N/2}}{\Gamma(1 + 3N/2)} (2m)^{3N/2} E^{3N/2 - 1}. \tag{3.58}$$

(b) Show that, similarly, one can also write the phase volume as

$$\Omega = \omega_c * \Omega_k, \tag{3.59}$$

with

$$\omega_c(E) = \int d\mathbf{q}\,\delta[E - \phi(\mathbf{q})] \tag{3.60}$$

the configurational structure function, and

$$\Omega_k(E) = \frac{\pi^{3N/2}}{\Gamma(1 + 3N/2)} (2m)^{3N/2} E^{3N/2} \tag{3.61}$$

the kinetic volume. ∎

3.5 Some History

For ergodic systems, the quantity $\Omega(E, \lambda)$ is the *Adiabatic Invariant*. This property was the cornerstone of the construction of the mechanical expression of entropy by Paul Hertz [3], which leads, in fact, to the same result that we have obtained, namely

$S = \ln \Omega$. The work of Hertz was later praised by Einstein [4], and was adopted by some textbooks of the German tradition, e.g., [5] and [6]. The first to put forward the expression $S = \ln \Omega$, was however Gibbs [7], but a closely related expression was in fact already implicit in Boltzmann's first paper [8], which was later picked up and further developed by Helmholtz [9, 10] and Boltzmann himself [11], see also the accounts by Gallavotti [12] and Darrigol [13], and the exercises below. The mechanical approach to expressing entropy of isolated systems remained however almost hidden and forgotten in the literature in favour of the celebrated Boltzmann combinatorial approach, dated 1877 [14].

3.5.1 Boltzmann Expressions of Microcanonical Entropy

In the 1866 paper titled "Über die mechanische Bedeutung des zweiten Hauptsatzes der Wärmetheorie" ("On the mechanical meaning of the second principle of the heat theory"), Boltzmann expresses the entropy associated to a single degree of freedom as the logarithm of the integrated kinetic energy [8]

$$S = \ln \int_0^\tau dt \, K(t), \qquad (3.62)$$

where τ is the recurrence time.

Exercise 3.4 (a) Show that this is equivalent to the expression in Eq. (2.22). (b) Using the equipartition theorem, prove that, in the many-body interacting case, assuming ergodicity, the expression (3.62) with K denoting the total kinetic energy, is equivalent to $S = \ln \Omega$ (i.e., at most they differ by a constant term). ∎

In the 1884 paper titled "Über die Eigenschaften monocyklischer und anderer damit verwandter Systeme" [11] ("On the properties of monocyclic and other related systems") Boltzmann finds (see e.g., Darrigol [13]) an expression for the entropy of a system with s degrees of freedom in the microcanonical ensemble which with our language and notation would read:

$$S(E, \lambda) = \ln \int d\mathbf{q} [E - \varphi(\mathbf{q}; \lambda)]_+^{s/2} \qquad (3.63)$$

where the symbol $[x]_+$ denotes x for $x \geq 0$, and 0 for $x < 0$.

Exercise 3.5 Use Eq. (3.49) to prove that Eq. (3.63) is equivalent to Eq. (3.9), i.e., at most they differ by a constant term. ∎

Note that the Boltzmann expression (3.63) reduces to Eq. (2.22) for $s = 1$.

References

1. Lebowitz, J.L., Penrose, O.: Phys. Today **26**(2), 23 (1973). https://doi.org/10.1063/1.3127948
2. Khinchin, A.: Mathematical Foundations of Statistical Mechanics. Dover, New York (1949)
3. Hertz, P.: Ann. Phys. (Leipzig) **338**, 225 (1910). https://doi.org/10.1002/andp.19103381202
4. Einstein, A.: Ann. Phys. (Leipzig) **339**, 175 (1911). https://doi.org/10.1002/andp.19113390111
5. Becker, R.: Theory of Heat. Springer, New York (1967)
6. Münster, A.: Statistical Thermodynamics, (vol. 1). Springer, Berlin (1969)
7. Gibbs, J.: Elementary Principles in Statistical Mechanics. Yale University Press, New Haven (1902)
8. Boltzmann, L.: Wiener Berichte **53**, 195 (1866)
9. von Helmholtz, H.: Journal für die reine und angewandte Mathematik **97**, 111 (1884). https://doi.org/10.1515/9783112342169-008
10. von Helmholtz, H.: Journal für die reine und angewandte Mathematik **97**, 317 (1884). https://doi.org/10.1515/9783112342169-019
11. Boltzmann, L.: Journal für die reine und angewandte Mathematik **98**, 68 (1885). https://doi.org/10.1515/crll.1885.98.68
12. Gallavotti, G.: Statistical Mechanics: A Short Treatise. Springer, Berlin (1999)
13. Darrigol, O.: Atoms, Mechanics, and Probability: Ludwig Boltzmann's Statistico-Mechanical Writings - An Exegesis. OUP Oxford (2018)
14. Boltzmann, L.: Sitzungsber. d. k. Akad. der Wissenschaften zu Wien **II 76**, 428 (1877)

Chapter 4
The Canonical Ensemble

4.1 A System in Weak Interaction with a Large Ideal Gas

We consider a system in interaction with an ideal gas composed of N particles. Let

$$H_{tot}(\mathbf{z}, \mathbf{w}) = H(\mathbf{z}) + H_{gas}(\mathbf{w}) + H_{int}(\mathbf{z}, \mathbf{w}) \tag{4.1}$$

where, $\mathbf{z}, \mathbf{w}$ are the system and gas phase space coordinates, respectively. H and H_{gas}, are their respective Hamiltonians, and $H_{int}(\mathbf{z}, \mathbf{w})$ is an interaction term that allows system and gas to exchange energy. Figure 4.1 exemplifies schematically one possible realisation of the physical scenario under consideration. We shall assume $H_{int}(\mathbf{z}, \mathbf{w})$ is such that the total Hamiltonian is ergodic at some energy E, and that for the calculations that follow the term $H_{int}(\mathbf{z}, \mathbf{w})$ is negligible. We shall comment later on this assumption. Under the ergodic hypothesis the probability density to find the composite system at $(\mathbf{z}, \mathbf{w})$ is given by the microcanonical distribution

$$\rho_\mu(\mathbf{z}, \mathbf{w}; E) = \frac{\delta[E - H(\mathbf{z}) - H_{gas}(\mathbf{w})]}{\omega_{tot}(E)}, \tag{4.2}$$

where we have neglected the interaction term. Using Eq. (3.44) the probability density of finding the system at $\mathbf{z}$ is

$$p(\mathbf{z}) = \frac{\omega_{gas}(E - H(\mathbf{z}))}{\omega_{tot}(E)}, \tag{4.3}$$

where ω_{gas} and ω_{tot} are the gas and total system structure function, respectively.

Using now Eq. (3.51) for the gas structure function we obtain:

$$p(\mathbf{z}) = \frac{\left[1 - \frac{H(\mathbf{z})}{E}\right]^{C_V - 1}}{\text{normalisation}}, \tag{4.4}$$

M. Campisi, *Lectures on the Mechanical Foundations of Thermodynamics*, SpringerBriefs in Physics, https://doi.org/10.1007/978-3-030-87163-5_4

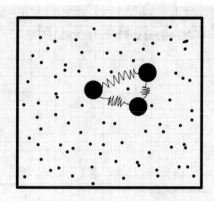

where $C_V = 3N/2$ is the gas heat capacity. Thus the probability distribution for the system is a power-law with an exponent given by the heat capacity of the gas.

Let us now introduce the parameter:

$$\theta = \frac{E}{C_V}, \tag{4.5}$$

and let's taking the limit of a very large gas, $C_V \to \infty$, while keeping θ constant. Using the formula:

$$\lim_{n \to \infty} \left(1 + \frac{x}{n}\right)^n = e^x, \tag{4.6}$$

we obtain

$$p(\mathbf{z}) \to \frac{e^{-H(\mathbf{z})/\theta}}{Z(\theta)}, \tag{4.7}$$

where

$$Z = \int d\mathbf{z}\, e^{-H(\mathbf{z})/\theta}. \tag{4.8}$$

The distribution $e^{-H(\mathbf{z})/\theta}/Z$ is called canonical distribution, and Z, its normalisation, is called the partition function.

One might wonder what is the physical meaning of θ. Since $E = (3N/2)\theta$, θ is the temperature that the gas would have if all its energy were fixed and equal to E. Note that the gas energy is not fixed, because it is in continuous interaction with the system. However, if the gas is enormously much bigger than the system, one may intuitively expect that practically all the energy, apart from negligible fluctuations is taken by the gas. Thus, one might conclude that θ represents the common temperature of the two systems. This statement can in fact be made rigorous thanks to the following

Theorem 4.1 (Canonical Equipartition Theorem)

$$\left\langle z_i \frac{\partial H}{\partial z_i} \right\rangle_c = \frac{1}{\beta}, \tag{4.9}$$

where z_i is any of the q's or the p's of the system, and the symbol $\langle \cdot \rangle_c$ denotes the average over the canonical distribution.

Definition 4.1 (*Canonical distribution*)

$$\rho_c(\mathbf{z}; \beta, \lambda) = \frac{e^{-\beta H(\mathbf{z}, \lambda)}}{Z(\beta, \lambda)}, \tag{4.10}$$

$$Z(\beta, \lambda) = \int d\mathbf{z} e^{-\beta H(\mathbf{z}, \lambda)}, \tag{4.11}$$

where we have written explicitly the dependence of the system Hamiltonian on an external parameter, and we used the standard notation $1/\beta$ instead of θ.

Theorem 4.1 tells indeed that θ has the meaning of common temperature of system and gas. To see that consider the total system. The microcanonical equipartition theorem tells us that the temperature is $\left\langle z_i \frac{\partial H}{\partial z_i} \right\rangle_\mu$. Integrating out the gas degrees of freedom, we have, in the $C_V \to \infty$ limit, $\left\langle z_i \frac{\partial H}{\partial z_i} \right\rangle_\mu \to \left\langle z_i \frac{\partial H}{\partial z_i} \right\rangle_c = 1/\beta = \theta$. Hence θ is the common temperature of system and gas. Based on this argument, it should not surprise that in the canonical ensemble, the average of $z_i \frac{\partial H}{\partial z_i}$ is one and the same for all degrees, just like in the microcanonical ensemble.

Let us prove Theorem 4.1:

$$\begin{aligned}
\left\langle z_i \frac{\partial H}{\partial z_i} \right\rangle_c &= \frac{1}{Z(\beta, \lambda)} \int d\mathbf{z} z_i \frac{\partial H}{\partial z_i} e^{-\beta H(\mathbf{z}, \lambda)} \\
&= -\frac{1}{\beta Z(\beta, \lambda)} \int d\mathbf{z} z_i \frac{\partial}{\partial z_i} e^{-\beta H(\mathbf{z}, \lambda)} \\
&= -\frac{1}{\beta Z(\beta, \lambda)} \left(\int d\mathbf{z}^i \left[z_i e^{-\beta H(\mathbf{z}, \lambda)} \right]_{-\infty}^{+\infty} - \int d\mathbf{z} e^{-\beta H(\mathbf{z}, \lambda)} \right) \\
&= \frac{1}{\beta}
\end{aligned}$$

where $\mathbf{z}^i = (z_0 \ldots z_{i-1}, z_{i+1} \ldots z_{2s})$, we have integrated by parts in z_i in the third line, and have assumed the Hamiltonian H grows indefinitely and fast enough as $|z_i|$ grows, that the term $\left[z_i e^{-\beta H(\mathbf{z}, \lambda)} \right]_{-\infty}^{+\infty}$ vanishes. This happens, for example, if the system is confined in a box potential.

4.2   Validity of the Heat Theorem within the Canonical Ensemble

Let us imagine we have a system that is distributed according to the canonical distribution. This allows us to calculate the following expectation values

$$T(\beta, \lambda) \doteq \left\langle z_i \frac{\partial H}{\partial z_i} \right\rangle_c = \frac{1}{\beta}, \tag{4.12}$$

$$\Lambda(\beta, \lambda) \doteq -\left\langle \frac{\partial H}{\partial \lambda} \right\rangle_c, \tag{4.13}$$

$$U(\beta, \lambda) \doteq \langle H \rangle_c, \tag{4.14}$$

representing temperature, force conjugated to λ (e.g. pressure when λ is the volume), and internal energy. We can now construct the differential form

$$\frac{dU(\beta, \lambda) + \Lambda(\beta, \lambda)d\lambda}{T(\beta, \lambda)}, \tag{4.15}$$

and ask the nontrivial question whether that differential is exact. If that is the case it means the canonical ensemble reproduces one of the central properties of thermodynamics (namely the validity of the heat theorem, or the conservativeness of the thermodynamic field), and may accordingly be considered as providing a good model of thermodynamics, just like the microcanonical ensemble.

The answer is given by the following canonical version of the generalised Helmholtz theorem

Theorem 4.2 *The differential in Eq. (4.15) is exact, namely there exist a function* $S(\beta, \lambda)$ *such that*

$$dS(\beta, \lambda) = \frac{dU(\beta, \lambda) + \Lambda(\beta, \lambda)d\lambda}{T(\beta, \lambda)}. \tag{4.16}$$

Furthermore, S is given by

$$S(\beta, \lambda) = \beta U(\beta, \lambda) + \ln Z(\beta, \lambda). \tag{4.17}$$

The proof is based on the following crucial relations:

$$U(\beta, \lambda) = -\frac{\partial}{\partial \beta} \ln Z, \tag{4.18}$$

$$\beta \Lambda(\beta, \lambda) = \frac{\partial}{\partial \lambda} \ln Z, \tag{4.19}$$

which the reader can check independently. The proof then easily follows by noticing that $dU = d\beta \partial U / \partial \beta + d\lambda \partial U / \partial \lambda$.

Exercise 4.1 Complete the proof of the theorem.   ∎

Note that S can be identified with the entropy of the system. Also note that $\ln Z = S - \beta U$ and its partial derivatives give $-U$ and $\beta \Lambda$. By inspection, see Sect. 1.2, we recognise that, quite surprisingly, $\ln Z$ can be identified with the Massieu Potential Ψ_F:

$$\Psi_F(\beta, \lambda) = \ln Z(\beta, \lambda), \tag{4.20}$$

which immediately gives us the free energy as

$$F(T, \lambda) = -T \ln Z(1/T, \lambda). \tag{4.21}$$

4.3 Properties of the Partition Function

Using Eq. (3.37) we see that:

$$Z(\beta, \lambda) = \int_0^\infty dE e^{-\beta E} \omega(E, \lambda) = \mathscr{L}[\omega(\cdot, \lambda)](\beta), \tag{4.22}$$

where we used the symbol $\mathscr{L}[\omega(\cdot, \lambda)](\beta)$ to denote the Laplace transform of ω with respect to its first argument, taken at β.

The composition rule for non-interacting systems follows immediately from the composition rule for the structure functions, Eq. (3.40):

$$Z(\beta, \lambda) = Z_1(\beta, \lambda) Z_2(\beta, \lambda), \tag{4.23}$$

where Z_i is the partition function of subsystem i. This relation can also be checked directly from the definition with the condition $H = H_1 + H_2$.

Since Ω is the integral in energy of ω, due to the well known properties of the Laplace transform, we have

$$Z(\beta, \lambda) = \beta \mathscr{L}[\Omega(\cdot, \lambda)](\beta). \tag{4.24}$$

4.4 Ideal Gas

The Hamiltonian of N free particles confined into a box of volume V with hard boundaries reads:

$$H = \sum_{i=1}^N H_i(\mathbf{q}_i, \mathbf{p}_i; V) = \sum_{i=1}^N \left[\frac{\mathbf{p}_i^2}{2m} + U_{\text{box}}(\mathbf{q}_i; V) \right], \tag{4.25}$$

where $(\mathbf{q}_i, \mathbf{p}_i)$ is the 6 dimensional phase space of particle i. Note that all single particle Hamiltonians H_i are identical. Let

$$Z_1(\beta, V) = \int d\mathbf{q}d\mathbf{p}e^{-\beta H_1(\mathbf{q},\mathbf{p})}.$$ (4.26)

Using the composition rule, it is then

$$Z(\beta, V) = Z_1(\beta, V)^N.$$ (4.27)

Now, Z_1 is the Laplace transform of the structure function of an ideal gas made up of $N = 1$ particles in 3d. Using Eq. (3.51), that equals:

$$\omega_1(E, V) = \frac{3}{2}\frac{\pi^{3/2}}{\Gamma(1+3/2)}(2m)^{3/2}VE^{3/2-1}.$$ (4.28)

Using the formula $\mathscr{L}[E^{1/2}] = \frac{\sqrt{\pi}}{2\beta^{3/2}}$, the according partition function reads:

$$Z_1(\beta, V) = \frac{3}{2}\frac{(2m\pi)^{3/2}}{\Gamma(1+3/2)}\frac{\sqrt{\pi}}{2}\frac{V}{\beta^{3/2}}.$$ (4.29)

Raising to the power N, we get

$$Z(\beta, V) = (2m\pi)^{3N/2}\frac{V^N}{\beta^{3N/2}}.$$ (4.30)

Using Eqs. (4.18), (4.19) we get:

$$U(\beta, V) = \frac{3N}{2\beta}, \quad \text{i.e.,} \quad U = \frac{3N}{2}T,$$ (4.31)

$$\beta P(\beta, V) = \frac{N}{V}, \quad \text{i.e.,} \quad PV = NT,$$ (4.32)

namely the thermodynamics of the ideal gas. The heat capacity at constant volume C_V is generally given by the formula:

$$C_V = -\beta^2\frac{\partial U}{\partial \beta},$$ (4.33)

which, in this specific case amounts, as expected, to $C_V = 3N/2$.

We see that, regardless of whether one uses the canonical or microcanonical formalism, one obtains one and the same equations of state for the ideal gas. We shall come back to this when discussing the topic of equivalence between ensembles.

4.5 Linear Response Coefficients and Fluctuations

The first derivatives of the Massieu potential, $\Psi_F = \ln Z$, give the average energy U and the average "force" $\beta \Lambda$ conjugated entropically to the parameter λ.

The second derivatives are accordingly linked to the so called linear response coefficients, e.g. , the heat capacity, the compressibility, or the electric and magnetic susceptibilities, according to the meaning of λ. For instance, see Eq. (4.33),

$$\frac{\partial^2 \Psi_F}{\partial \beta^2} = -\frac{\partial U}{\partial \beta} = \beta^{-2} C_V. \tag{4.34}$$

On the other hand, using the Definition 4.1, it is not difficult to see that:

$$\frac{\partial^2 \Psi_F}{\partial \beta^2} = \langle H^2 \rangle - \langle H \rangle^2. \tag{4.35}$$

We thus obtain the crucial result:

$$\langle H^2 \rangle - \langle H \rangle^2 = \beta^{-2} C_V = T^2 C_V. \tag{4.36}$$

This is an instance of the so-called fluctuation dissipation theorem, linking responses (C_V tells you how much the average energy changes as you change temperature) to fluctuations. Note the important fact that, for an ideal gas, C_V is proportional to N, that implies that the relative fluctuations $\sqrt{(\langle H^2 \rangle - \langle H \rangle^2)/\langle H \rangle^2}$ vanish as $1/\sqrt{N}$ in the large N limit.

The second crucial observation is that, since the left hand side of Eq. (4.36) is non-negative, it is $C_V \geq 0$, in the canonical ensemble. Note that we do not have any such condition in the microcanonical ensemble, see Sect. 2.5.3.

It is interesting to take the second derivative of Ψ_F with respect to λ. We obtain the relation:

$$\left\langle \left(\frac{\partial H}{\partial \lambda}\right)^2 \right\rangle - \left\langle \frac{\partial H}{\partial \lambda} \right\rangle^2 = \frac{1}{\beta} \left[\frac{\partial \Lambda}{\partial \lambda} + \left\langle \frac{\partial^2 H}{\partial \lambda^2} \right\rangle \right]. \tag{4.37}$$

Take as example a magnetic system. The Hamiltonian reads $H = H_0 - M\mathcal{H}$, where M is the system magnetic moment, and $\mathcal{H}$ is the external field (for simplicity consider an isotropic system and the field as pointing in the positive x direction). The field intensity $\mathcal{H}$ plays the role of λ, and the conjugate force Λ is the average magnetisation $\langle M \rangle$ in the direction x. In this case $\partial^2 H / \partial \mathcal{H}^2 = 0$, and the above equation becomes:

$$\langle M^2 \rangle - \langle M \rangle^2 = \frac{1}{\beta} \frac{\partial \langle M \rangle}{\partial \mathcal{H}} = T \chi_m \tag{4.38}$$

where $\chi_m = \partial\langle M\rangle/\partial\mathcal{H}$ is the magnetic susceptibility, which should clearly be non-negative.

One might as well consider the cross derivative:

$$-\frac{\partial\langle H\rangle}{\partial\lambda} = \frac{\partial^2\Psi_F}{\partial\lambda\partial\beta} = -\left\langle\frac{\partial H}{\partial\lambda}\right\rangle + \beta\left\langle H\frac{\partial H}{\partial\lambda}\right\rangle - \beta\langle H\rangle\left\langle\frac{\partial H}{\partial\lambda}\right\rangle, \tag{4.39}$$

which provides a relation for the cross correlation function of H with $\partial H/\partial\lambda$

$$\left\langle H\frac{\partial H}{\partial\lambda}\right\rangle - \langle H\rangle\left\langle\frac{\partial H}{\partial\lambda}\right\rangle = T\left[\left\langle\frac{\partial H}{\partial\lambda}\right\rangle - \frac{\partial\langle H\rangle}{\partial\lambda}\right] \tag{4.40}$$

Exercise 4.2 Prove Eqs. (4.35), (4.37) and (4.40).   ∎

Exercise 4.3 Consider the case of a single particle gas confined in a box potential of volume L in 1 dimension. L plays the role of λ, and Λ is the pressure P, accordingly. The left hand side of Eq. (4.37) represents the fluctuation of the force exerted by the system on the walls of the container. The first term in the right hand side of (4.37) is related to the isothermal compressibility $\chi_T = -(1/L)(\partial L/\partial P)$. (a) Modelling the box potential as a limit of a succession of smooth potentials, evaluate the second term on the r.h.s. of Eq. (4.37). (b) Generalise to a gas of N particles in 3D.   ∎

4.6 Finite Bath Ensemble

The distribution in Eq. (4.4) is sometimes referred to in the literature, as the Finite Bath Ensemble [1]. The Finite Bath Ensemble can be written in the following meaningful form [1]:

$$\rho_C(\mathbf{z}; U, \lambda) = \frac{1}{N_C(U, \lambda)}\left[1 - \frac{H(\mathbf{z}; \lambda) - U}{CT(U, \lambda)}\right]_+^{C-1} \tag{4.41}$$

where C is a short notation for C_V, $N_C(U, \lambda)$ is the normalisation, $[x]_+ = x\theta(x)$ and $T(U, V)$ is the solution (which is assumed to be unique) of

$$\frac{\int d\mathbf{z}\left[1 - \frac{H(\mathbf{z};\lambda)-U}{CT}\right]_+^{C-1} H(\mathbf{z}; \lambda)}{\int d\mathbf{z}\left[1 - \frac{H(\mathbf{z};\lambda)-U}{CT}\right]_+^{C-1}} = U. \tag{4.42}$$

Exercise 4.4 (a) Prove Eq. (4.41). Hint. Note that U is the average energy of the system, and CT is the average energy of the bath, therefore $E = U + CT$. (b) Prove that the Equipartition Theorem is obeyed in the Finite Heat Bath Ensemble:

$$\left\langle z_i \frac{\partial H}{\partial z_i} \right\rangle_C = T(U; \lambda), \tag{4.43}$$

where $\langle \cdot \rangle_C$ denotes average over ρ_C, Eq. (4.41). (c) Prove that the Heat Theorem holds in the Finite Heat Bath Ensemble, and write down the according expression of entropy $S_C(U, \lambda)$. (d) Prove that the ensembles ρ_C interpolate between canonical and microcanonical ensembles, namely, that ρ_C tends to the canonical ensemble in the limit $C \to \infty$, and it tends to the microcanonical ensemble in the limit $C \to 0$. Show that the entropy $S_C(U, \lambda)$ reduces to the canonical and microcanonical expressions of entropy, in the according limits. (e) Evaluate the energy fluctuations in the Finite Bath Ensemble. (f) Derive the equation of state for an ideal gas in contact with a finite heat bath. How does it compare with the according equations obtained in the canonical and microcanonical ensembles? (This problem is treated in [1]). ∎

4.7 A System in Weak Contact with a Logarithmic Oscillator

In Sect. 2.5.2 we have seen that the structure function of a logarithmic oscillator is exponential in energy, $\omega(E) \propto e^{E/T}$. If a system with Hamiltonian $H(\mathbf{z})$ stays in weak contact with a logarithmic oscillator, then, according to Eq. (3.44) its phase space probability density obeys:

$$p(\mathbf{z}; E) \propto e^{[E - H(\mathbf{z})]/T}, \tag{4.44}$$

with E the total system+log-oscillator energy. By normalising the expression above one gets:

$$p(\mathbf{z}) = \frac{e^{-H(\mathbf{z})/T}}{\int d\mathbf{z} e^{-H(\mathbf{z})/T}} . \tag{4.45}$$

This says that a system in weak contact with a logarithmic oscillator samples the canonical distribution, provided the total system dynamics are ergodic, which ensures the applicability of Eq. (3.44). From the physical point of view that is because the logarithmic oscillator has infinite heat capacity, just like an infinite thermal reservoir.

Figure 4.2 shows the results of numerical simulations of a 1D system composed of two particles in a box, making hard-core collisions with a particle confined to a logarithmic potential [2]. Since the structure function for two hard spheres is a constant (does not depend on energy), the canonical probability density function in energy reads in this specific case, see Eq. (3.45)

$$P(E_S) = \frac{e^{-E_S/T}}{\int dE_S e^{-E_S/T}} . \tag{4.46}$$

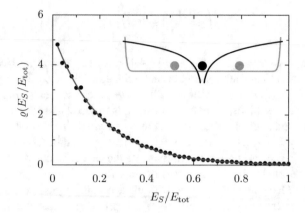

Fig. 4.2 Probability density function of energy for a system of two particles in a 1D box performing short ranged collisions with a log-oscillator. Red line: Gibbs distribution at temperature T. The total system is schematically represented in the inset with the system of interest (two orange particles) confined to the box potential (orange curve), and the log-oscillator (black particle) confined to the logarithmic potential (black curve). Reprinted with permission from [2]. Copyright 2012 by the American Physical Society

Note the agreement between theory and numerical experiment.

Exercise 4.5 Show that the structure function for two free particles in a box is a constant (does not depend on energy). ■

We have just proved the a logarithmic oscillator, when allowed to weakly interact with a mechanical systems, may act as a "thermostat", namely, as a mechanism that makes the system sample the canonical distribution at a certain temperature. Devising such "thermostatting" mechanisms is, for obvious reasons, of great relevance in computational physics, and there is a good amount of literature devoted to the problem of generating molecular dynamics at constant temperature. Various thermostats have been proposed and are routinely employed to simulate systems evolving in a fixed temperature environment. Among them are stochastic methods and deterministic methods. Among the deterministic ones the method of Nosé [3], is closely related to the above method, as it uses as well a log-oscillator, but it couples it to the main system in a very peculiar way. The interested reader may find a review of various methods to generate fixed temperature dynamics, e.g., in Ref. [4].

References

1. Campisi, M., Talkner, P., Hänggi, P.: Phys. Rev. E **80**, 031145 (2009). https://doi.org/10.1103/PhysRevE.80.031145
2. Campisi, M., Zhan, F., Talkner, P., Hänggi, P.: Phys. Rev. Lett. **108**, 250601 (2012). https://doi.org/10.1103/PhysRevLett.108.250601

3. Nosé, S.: J. Chem. Phys. **81**, 511 (1984). https://doi.org/10.1063/1.447334
4. Hünenberger, P.H : Thermostat Algorithms for Molecular Dynamics Simulations. In: Dr. Holm C., Prof. Dr. Kremer K. (eds) Advanced Computer Simulation. Advances in Polymer Science, vol 173, pp. 105–149. Springer, Berlin, Heidelberg (2005). https://doi.org/10.1007/b99427

Chapter 5
The TP Ensemble

5.1 A System in Contact with a Thermal Reservoir and Subject to a Constant Pressure

The microcanonical ensemble is an ensemble with fixed energy and external parameter, typically the volume. For this reason it is also called the EV ensemble. In order to get the canonical ensemble, we have considered our system as immersed in a thermal bath that fixes its temperature, while its energy can fluctuate. Temperature and volume then parametrise the canonical ensemble. For this reason it is also known as the TV ensemble (or βV if you prefer).

We consider now the case where the system of interest is in contact with a thermal bath that fixes its temperature and is subject to a constant applied pressure. This latter situation can be achieved by allowing the volume of the box that contains the system to fluctuate while an external constant pressure is applied to its walls. Figure 5.1 shows how that can be obtained in practice. One diathermal wall of the box allows your system to exchange energy with a reservoir at temperature T. Another wall is a movable piston of some area A on which a constant force $F = PA$, is applied, so that a constant pressure P is imposed on the system. In this scenario the volume V of the vessel is no longer an external parameter: it is a dynamical variable. Let X be the position of the piston, p_X the according conjugate momentum, and M its mass. It is $V = AX$ and its canonically conjugated quantity is $p_V = p_X/A$, such that Vp_V has the dimension of action. The extended Hamiltonian describing system and piston reads accordingly:

$$H_{\text{ext}}(\mathbf{z}, V, p_V) = H_S(\mathbf{z}) + h(\mathbf{z}, V) + A^2 \frac{p_V^2}{2M} + PV \tag{5.1}$$

where the interaction between system and piston $h(\mathbf{z}, V)$ is left unspecified, and $H_S(\mathbf{z})$ is the system Hamiltonian. Note the $+$ sign for the PV term. That expresses the fact that the force is directed in the opposite direction as that of growing volume.

© The Author(s), under exclusive license to Springer Nature Switzerland AG 2021
M. Campisi, *Lectures on the Mechanical Foundations of Thermodynamics*,
SpringerBriefs in Physics,
https://doi.org/10.1007/978-3-030-87163-5_5

Fig. 5.1 A thermostated
system subject to a constant
pressure

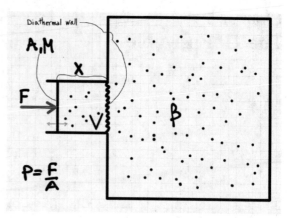

Introducing the so called "inclusive" Hamiltonian[1]

$$H(\mathbf{z}, V) = H_S(\mathbf{z}) + h(\mathbf{z}, V), \tag{5.2}$$

and using the fact that the extended system is thermostatted at temperature $T = 1/\beta$,
the probability density of finding the extended system at $(\mathbf{z}, V, p_V)$ reads:

$$\rho(\mathbf{z}, V, p_V) = \frac{e^{-\beta A^2 p_V^2/(2M)} e^{-\beta H(\mathbf{z},V)} e^{-\beta PV}}{\int dV \, dp_V \, d\mathbf{z} \, e^{-\beta A^2 p_V^2/(2M)} e^{-\beta H(\mathbf{z},V)} e^{-\beta PV}}. \tag{5.3}$$

Since the distribution factorises into the product of a function of p_V times a function of
$\mathbf{z}, V$, the probability density of finding the system at $\mathbf{z}$ and piston at V is immediately
obtained as:

$$\rho(\mathbf{z}, V) = \frac{e^{-\beta H(\mathbf{z},V)} e^{-\beta PV}}{\int dV \, d\mathbf{z} \, e^{-\beta H(\mathbf{z},V)} e^{-\beta PV}}. \tag{5.4}$$

Note that P enters in the above equations always along with the multiplying factor
β. It is thus natural (and also convenient for practical caluclations), to parametrise
the so-called isothermal-isobaric ensemble with the parameters β and the quantity
π bearing the meaning of pressure over temperature βP.

---

[1] This terminology is dictated by the need to clearly distinguish between the bare system Hamiltonian
H_S, which does not include the interaction with the piston, and the Hamiltonian H that instead
includes it [1, 2].

Definition 5.1 (TP-ensemble)

$$\rho(\mathbf{z}, V; \beta, \pi) = \frac{e^{-\pi V} e^{-\beta H(\mathbf{z}; V)}}{Q(\beta, \pi)}, \tag{5.5}$$

$$Q(\beta, \pi) = \int_0^\infty dV \, e^{-\pi V} Z(\beta, V). \tag{5.6}$$

Here $Q(\beta, \pi)$ is the new partition function, and we have explicitly expressed the quantities β, π that parametrise the ensemble.

Note how, just like the canonical partition function $Z(\beta, V)$ is the Laplace transform of the microcanonical density of states $\omega(E, V)$ with respect to the exchange of E with its entropic partner β, so is the TP partition function $Q(\beta, \pi)$ the Laplace transform of the canonical partition function $Z(\beta, V)$, with respect to the exchange of V with its entropic partner π.

5.2 Validity of the Heat Theorem within the TP Ensemble

First of all we check that β and π have indeed the meaning of inverse temperature and inverse temperature times pressure within this ensemble. The answer is yes to both questions.

The answer to the first question is given by the

Theorem 5.1 (TP Equipartition Theorem)

$$\left\langle z_i \frac{\partial H}{\partial z_i} \right\rangle_{\beta, \pi} = \frac{1}{\beta} \tag{5.7}$$

where z_i is any of the q's or the p's of the system and $\langle \cdot \rangle_{\beta,\pi}$ denotes average over the TP ensemble.[2]

Exercise 5.1 Prove Theorem 5.1. ∎

The answer to the second question is given by the relation

$$-\left\langle \frac{\partial H}{\partial V} \right\rangle_{\beta, \pi} = \frac{\pi}{\beta}, \tag{5.8}$$

saying that π is indeed the physical pressure times β. Since in our construction $\pi = \beta P$, with P being the externally applied pressure, the above equation says that the applied "external" pressure equals the "internal pressure" of our system. Namely, the system stays in equilibrium.

---

[2] Note that in fact, thanks to the canonical partition theorem it also holds $\langle V \partial H / \partial V \rangle = \langle p_V^2 / (2m) \rangle = 1/\beta$.

Exercise 5.2 Prove Eq. (5.8). Hint: integrate by parts in V, and use the fact that the support of $H(\mathbf{z}; V)$ reduces to a point (i.e., a zero measure set) when the volume of the confining box V tends to zero.   ∎

Now that we have assigned physical meaning to β and π, we can, as usual define the quantities:

$$T(\beta, \pi) \doteq \left\langle z_i \frac{\partial H}{\partial z_i} \right\rangle_{\beta, \pi} = \frac{1}{\beta}, \tag{5.9}$$

$$P(\beta, \pi) \doteq -\left\langle \frac{\partial H}{\partial V} \right\rangle_{\beta, \pi} = \frac{\pi}{\beta}, \tag{5.10}$$

$$U(\beta, \pi) \doteq \langle H \rangle_{\beta, \pi}, \tag{5.11}$$

$$\bar{V}(\beta, \pi) \doteq \langle V \rangle_{\beta, \pi}. \tag{5.12}$$

Note that V is now fluctuating, therefore we have introduced the average volume $\bar{V}(\beta, \pi)$. We can now ask ourselves the nontrivial question wether the quantity representing the heat differential over temperature

$$\frac{dU(\beta, \pi) + P(\beta, \pi)d\bar{V}(\beta, \pi)}{T(\beta, \pi)} \tag{5.13}$$

is exact. The answer is given by the following:

Theorem 5.2 *The differential in Eq. (5.13) is exact, namely there exist a function $S(\beta, \pi)$ such that*

$$dS(\beta, \pi) = \frac{dU(\beta, \pi) + P(\beta, \pi)d\bar{V}(\beta, \pi)}{T(\beta, \pi)}. \tag{5.14}$$

Furthermore, S is given by

$$S(\beta, \pi) = \beta U(\beta, \pi) + \pi \bar{V}(\beta, \pi) + \ln Q(\beta, \pi). \tag{5.15}$$

The proof is based on the following crucial relations:

$$U(\beta, \pi) = -\frac{\partial}{\partial \beta} \ln Q(\beta, \pi), \tag{5.16}$$

$$\bar{V}(\beta, \pi) = -\frac{\partial}{\partial \pi} \ln Q(\beta, \pi). \tag{5.17}$$

Note the simplicity of these relations.

Exercise 5.3 Prove Eqs. (5.16 and 5.17) and complete the proof of the theorem. ∎

Note that S can be identified with the entropy of the system. Also note that $\ln Q = S - \beta U - \pi V$ and its partial derivatives give $-U$ and $-V$. By inspection, see Sect. 1.2, we recognise that $\ln Q$ can be identified with the Massieu Potential Ψ_G:

$$\Psi_G(\beta, \pi) = \ln Q(\beta, \pi), \qquad (5.18)$$

which immediately gives us the Gibbs free energy as

$$G(T, P) = -T \ln Q(1/T, P/T). \qquad (5.19)$$

5.3 Fluctuations

Due to the exponential dependence of the distribution in both π and β, the evaluation of fluctuations and correlation functions is particularly easy in the TP-ensemble: We have:

$$\frac{\partial^2 \Psi_G}{\partial \beta^2} = \langle H^2 \rangle - \langle H \rangle^2 = -\frac{\partial U}{\partial \beta}, \qquad (5.20)$$

where one should remember that $U = U(\beta, \pi)$. Expressing the internal energy U as a function of T, P as in standard thermodynamics, we have[3]:

$$\langle H^2 \rangle - \langle H \rangle^2 = T^2 \left(\frac{\partial U}{\partial T} \right)_P + PT \left(\frac{\partial U}{\partial P} \right)_T. \qquad (5.23)$$

The second derivative of Ψ_G gives the volume fluctuations

$$\frac{\partial^2 \Psi_G}{\partial \pi^2} = \langle V^2 \rangle - \langle V \rangle^2 = -\frac{\partial V}{\partial \pi} = TV\chi_T, \qquad (5.24)$$

where $\chi_T = -(1/V)(\partial V/\partial P)_T$ is the isothermal compressibility.

---

[3] To see that define the new function $U^*(\beta, P) = U(\beta, \beta P)$. We have:

$$\left(\frac{\partial U}{\partial P} \right)_T \doteq \frac{\partial U^*}{\partial P} = \beta \frac{\partial U}{\partial (\beta P)}, \qquad (5.21)$$

$$\left(\frac{\partial U}{\partial \beta} \right)_P \doteq \frac{\partial U^*}{\partial \beta} = \frac{\partial U}{\partial \beta} + P \frac{\partial U}{\partial (\beta P)} = \frac{\partial U}{\partial \beta} + \frac{P}{\beta} \frac{\partial U^*}{\partial P} = \frac{\partial U}{\partial \beta} + PT \left(\frac{\partial U}{\partial P} \right)_T. \qquad (5.22)$$

When expressing quantities in terms of T instead of β, one similarly obtains $\partial/\partial \beta = -T^2 \partial/\partial T$.

The cross derivatives give the $H - V$ correlation function as:

$$\frac{\partial^2 \Psi_G}{\partial \pi \partial \beta} = \langle HV \rangle - \langle H \rangle \langle V \rangle = -\frac{\partial U}{\partial \pi} = -T \left(\frac{\partial U}{\partial P} \right)_T \tag{5.25}$$

$$= -\frac{\partial V}{\partial \beta} = T^2 \left(\frac{\partial V}{\partial T} \right)_P + PT \left(\frac{\partial V}{\partial P} \right)_T. \tag{5.26}$$

5.4   Ideal Gas

It is instructive to study the ideal gas in the TP ensemble. Taking the Laplace transform of the canonical partition function, Eq. (4.30), gives:

$$Q(\beta, \Pi) = N!(2m\pi)^{3N/2} \frac{1}{\beta^{3N/2}} \frac{1}{\Pi^{N+1}}, \tag{5.27}$$

where we used the capital letter Π for the entropic partner of V, in order to not make confusion with the real number π. Equations (5.16 and 5.17) give the ideal gas equations:

$$U(\beta, \Pi) = \frac{3N}{2\beta}, \tag{5.28}$$

$$\bar{V}(\beta, \Pi) = \frac{N+1}{\Pi}, \tag{5.29}$$

namely, $U = (3N/2)T$, and $P\bar{V} = (N+1)T$. One naturally wonders where the $+1$ term comes from. The answer is in our model: that comes from the extra degree of freedom associated to the fluctuating piston.

From Eq. (5.20) we get

$$\langle H^2 \rangle - \langle H \rangle^2 = \beta^{-2} \frac{3N}{2}. \tag{5.30}$$

Noting that for an ideal gas $C_V = 3N/2$ the energy fluctuations coincide with those obtained in the canonical ensemble. This is a peculiar coincidence that holds for the ideal gas, but not in general. From Eq. (5.24) we get:

$$\langle V^2 \rangle - \langle V \rangle^2 = \frac{N+1}{\Pi^2} = \frac{\langle V \rangle}{\Pi}. \tag{5.31}$$

Note that, in the limit of large N at fixed density $N/\bar{V}$, both the relative energy and volume fluctuations vanish.

From Eq. (5.25) we see that the energy of the system H and the volume of the vessel are uncorrelated for an ideal gas in the TP ensemble.

Fig. 5.2 An isolated system subject to a constant pressure

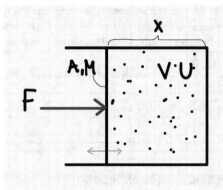

5.5 Thermally Insulated System Subject to a Constant Pressure

At this point of our discussion it becomes natural to ask ourselves about the ensemble that emerges when we fix energy and pressure, a topic that is not treated in standard textbooks. The physical scenario is that depicted in the Fig. 5.2. A constant force $F = PA$ is applied on the movable piston. The extended system is thermally insulated, and evolves according to the Hamiltonian (5.1), hence its total energy E is conserved. Accordingly, E and P can be used to parametrise the ensemble. Assuming ergodic dynamics, the extended system obeys the statistics:

$$\rho_{\text{ext}}(\mathbf{Z}; E, P) = \frac{\delta[E - H_{\text{ext}}(\mathbf{Z}; P)]}{\omega(E, P)}, \tag{5.32}$$

where $\mathbf{Z}$ is a short notation for the extended system phase point $\mathbf{z}$, V, p_V and $\omega(E, P) = \int d\mathbf{Z}\delta[E - H_{\text{ext}}(\mathbf{Z}; P)]$.

Integrating out the piston one obtains the probability density $\rho(\mathbf{z}; E, P)$ to find the system at $\mathbf{z}$, given that the total energy is E and the applied pressure is P.

Exercise 5.4 Show that $\rho(\mathbf{z}; E, P)$ can be written as:

$$\rho(\mathbf{z}; E, P) = \frac{\int dV[E - H(\mathbf{z}; V) - PV]_+^{-1/2}}{R(E, P)}, \tag{5.33}$$

$$R(E, P) = \int d\mathbf{z}dV[E - H(\mathbf{z}; V) - PV]_+^{-1/2}. \tag{5.34}$$

∎

According to the microcanonical equipartition theorem, it is $\langle z_i \partial H / \partial z_i \rangle = T(E, P)$, where $T(E, P) = \Omega(E, P) / \omega(E, P)$, with $\Omega(E, P) = \int d\mathbf{Z}\theta[E - H_{\text{ext}}(\mathbf{Z}; P)]$. In other words, introducing the function

$$S(E, P) = \log \Omega(E, P), \tag{5.35}$$

it is:

$$\frac{1}{T(E, P)} = \frac{\partial S}{\partial E}. \tag{5.36}$$

On the other hand it can be proved that:

$$P = -\left\langle \frac{\partial H}{\partial V} \right\rangle, \tag{5.37}$$

$$\frac{V(E, P)}{T(E, P)} = -\frac{\partial S}{\partial P}. \tag{5.38}$$

where $V(E, P) = \langle V \rangle$ denotes the average volume. The first of the above nicely tells that indeed P the pressure that is externally applied equals the average force that the main system exerts on the piston, confirming that the picture we are drawing is consistent.

Exercise 5.5 Prove Eqs. (5.37 and 5.38).   ∎

It follows that

$$dS = \frac{1}{T}dE - \frac{\langle V \rangle}{T}dP, \tag{5.39}$$

that is

$$dE = TdS + \langle V \rangle dP. \tag{5.40}$$

Now note that, neglecting the interaction energy h (assume, e.g., extremely short ranged wall-particles collisions), it is $E = \langle H \rangle + P\langle V \rangle + \langle P_V^2/(2M) \rangle = U(E, P) + PV(E, P) + 2T(E, P)$. Assuming as well a large main system, that is assuming $U \gg T$, it is $E = U + PV$, and therefore $dE = dU + PdV + VdP$. Combining, we have:

$$TdS = dU + PdV \tag{5.41}$$

By comparison, we see that S can be identified with the entropy.

References

1. Jarzynski, C.: C. R. Phys. **8**, 495 (2007). https://doi.org/10.1016/j.crhy.2007.04.010
2. Campisi, M., Hänggi, P., Talkner, P.: Rev. Mod. Phys. **83**(3), 771 (2011). https://doi.org/10.1103/RevModPhys.83.771

Chapter 6
The Grandcanonical Ensemble

6.1 Relaxing the Constraint on N

An ensemble of paramount importance in statistical mechanics is the (β, V, μ) ensemble, or grandcanonical ensemble, obtained from the canonical ensemble by relaxing the condition of fixed number of particles N, and fixing accordingly the chemical potential μ which determines its average value $\bar{N}$. The idea, see Fig. 6.1 is to have a very large system contained in a large box and to focus our interest only on a sub-region of volume V. The system of interest then is open and freely exchanges particles with the rest.

We are interested in finding an expression for the probability

$$\rho(N, \mathbf{z})d\mathbf{z} \tag{6.1}$$

to find N particles in the volume V *and* to find their phase coordinates within the element of measure $d\mathbf{z}$ around the phase point $\mathbf{z}$, given the temperature $1/\beta$ and the overall density of the gas ϱ. N is a stochastic, i.e., fluctuating quantity obeying some statistics $\mathcal{P}(N)$. We shall assume that the stochastic process that governs the statistics $\mathcal{P}(N)$ is a *spatial Poisson process*, that is we shall assume the following [1, 2]:

1. $N \in \mathbb{N}_+$, and $0 < \mathcal{P}(N) < 1$ for $V > 0$
2. $\mathcal{P}(N)$ depends only on the volume of the region V (not its shape), and $P(N \geq 1) \to 0$, for $V \to 0$
3. If $V_1, V_2 \ldots V_k$ denote the volumes of k disjoint regions inside the big box, and $N_1, N_2, \ldots N_k$ are the number of particles in each, then $N_1, N_2, \ldots N_k$, are mutually independent and $N = \sum_i N_i$
4. $\lim_{V \to 0} \frac{P(N \geq 1)}{P(N = 1)} = 1$

Some comments are in order. Regarding the first assumption, by definition our stochastic quantity N takes only non-negative integer values. Assuming that the process is *homogeneous in space*, implies that there are no forbidden regions, and that

Fig. 6.1 Physical scenario
for the grandcanonical
distribution. Particles can
freely enter and exit from the
region of interest of volume
V. The reservoir, which
should be understood as an
infinitely large system, fixes
the temperature and
chemical potential

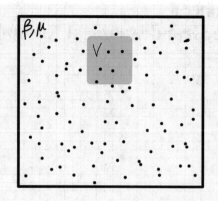

regardless of where we locate our region of interest it is $0 < \mathcal{P}(N) < 1$ for $V > 0$.
The second assumption is also a reflection of an assumed spatial homogeneity of
the process, and that, accordingly, for vanishing volume, the probability of finding
a particle therein also vanishes. The third assumption is basically an assumption of
lack of correlations between disjoint regions. This is plausible when the total system
contains an infinitely large number of particles which interact with very short range
forces: if there are few particles, knowing that a good amount of them is in one
region already tells you that few are in another, and the two numbers are then clearly
correlated. Similarly, if there are long range forces, e.g. attractive, the fact that there
is a good amount of particles in one region, influences the probability of finding
particles in the surrounding regions. The fourth assumption says that in the limit of
vanishing volume, finding some particles in the region has the same probability as
finding just one. That is plausible under the assumption that the particles occupy some
finite volume and cannot overlap, namely they interact via some hard-core repulsion.
A very large gas of hard-spheres making elastic collisions, would reasonably well
satisfy all four assumptions.

Under the above assumption, the statistics $\mathcal{P}(N)$ is of Poisson form [1, 2]:

$$\mathcal{P}(N) = \frac{e^{-f} f^N}{N!}, \tag{6.2}$$

where $f = \langle N \rangle$ is the average number of particles in the volume V. Given that there
are N particles in the volume V, the conditional probability density of finding them
at $\mathbf{z}$ is, according to the canonical prescription, given by:

$$\rho(\mathbf{z}|N)d\mathbf{z} = \frac{e^{-\beta H_N(\mathbf{z}; V)}}{Z(\beta, V, N)} d\mathbf{z}. \tag{6.3}$$

Here, the argument V in $H_N(\mathbf{z}; V)$ comes from a "fictitious" box potential that
confines the system to the volume V, which shall be included, in order to ensure that

the probability is zero if any of the particles is not inside the box. The joint probability $\rho(N, \mathbf{z})d\mathbf{z}$ is given, according to Bayes rule, by the product $\rho(\mathbf{z}|N)\mathcal{P}(N)d\mathbf{z}$:

$$\rho(N, \mathbf{z})d\mathbf{z} = \frac{f^N e^{-f}}{N!} \frac{e^{-\beta H_N(\mathbf{z}, V)}}{Z(\beta, V, N)} d\mathbf{z}. \tag{6.4}$$

We shall further assume that

$$Z(\beta, V, N) = \zeta(\beta, V)^N. \tag{6.5}$$

Note that this is the case of the ideal gas, see Eq. (4.27), and also, in the large N limit, the case of general systems with short range interactions. Introducing the quantity $e^\alpha = f/\zeta$ and noting that the $\mathbf{z}$ dependence is all contained in H_N, the probability $\rho(N, \mathbf{z})d\mathbf{z}$ can be equivalently written as

$$\rho(N, \mathbf{z})d\mathbf{z} = \frac{e^{\alpha N} e^{-\beta H_N(\mathbf{z}; V)} d\mathbf{z}/N!}{\sum_{K=1}^{\infty} e^{\alpha K} \int e^{-\beta H_K(\mathbf{z}; V)} d\mathbf{z}/K!} \tag{6.6}$$

We thus obtain an ensemble parametrised by β, V, α, namely the grandcanonical ensemble defined as follows:

Definition 6.1 (*grandcanonical ensemble*)

$$\rho(N, \mathbf{z}; \beta, V, \alpha)d\mathbf{z} = \frac{e^{\alpha N} e^{-\beta H_N(\mathbf{z}; V)}}{\Xi(\beta, V, \alpha)} \frac{d\mathbf{z}}{h^{3N} N!}, \tag{6.7}$$

$$\Xi(\beta, V, \alpha) = \sum_{N=1}^{\infty} e^{\alpha N} \int e^{-\beta H_K(\mathbf{z}; V)} \frac{d\mathbf{z}}{h^{3N} N!}. \tag{6.8}$$

Here, recalling the remarks in Sect. 3.1, we have multiplied and divided by the factor h^{3N}, where h is a constant with the units of action, in such a way that the grand-canonical partition function Ξ is dimensionless. We shall further comment on this below.

6.2 A Slight Detour: Gibbs "Correct" Counting

We recall that the microcanonical entropy $S(E, V, N) = \ln \Omega(E, V, N)$ is defined up to an additive constant $C(N)$, see Sect. 3.1. Accordingly, the phase volume Ω, Eq. (3.10), as well as the partition function Z, Eq. (4.11) is defined up to a multiplicative constant $K(N)$. Equation (6.8) suggests to chose the constant $K(N) = 1/(h^{3N} N!)$, namely to redefine the microcanonical phase volume, and accordingly the canonical partition function as

$$\Omega(E, V, N) = \int \frac{d\mathbf{z}}{h^{3N} N!} \theta[E - H_N(\mathbf{z}; V)], \tag{6.9}$$

$$Z(\beta, V, N) = \int \frac{d\mathbf{z}}{h^{3N} N!} e^{-\beta H_N(\mathbf{z}; V)}. \tag{6.10}$$

With this specific choice of the multiplicative constant, the dimensional consistency of entropy, free entropy and grand Massieu potential is obeyed, and also the grand-canonical partition function Ξ is the "discrete" Laplace transform of the canonical partition function Z, with respect to the change $N \leftrightarrow -\alpha$

$$\Xi(\beta, V, \alpha) = \sum e^{\alpha N} Z(\beta, V, N), \tag{6.11}$$

just like the partition function Z is the Laplace transform of the structure function ω, with respect to the change $E \leftrightarrow \beta$.[1]

Interestingly, the choice of multiplicative constant $K(N) \propto 1/N!$ is the one that solves the famous Gibbs paradox, concerning the mixing entropy of identical gases, which is discussed in most textbooks, see e.g. [3]. The prescription of rescaling the phase volume element $d\mathbf{z}$ by $h^{3N} N!$ is commonly referred to in the literature as the correct Gibbs counting. Quantum theory explains the emergence of the $N!$ based on the symmetry of the many-body wave functions of identical particles under permutation of the arguments, and also tells that h is Planck's constant. This topic is beyond the scope of the present set of lectures, the interested reader may find it discussed in many textbooks, e.g., [4]. Here we remark that, within the present approach the appearance of the $N!$ term is a reflection of the assumed Poissonian character of the statistics, hence it is not linked to an inherent property of distinguishability of the particles, rather to us ignoring their possible differences.[2]

---

[1] An issue however remains in regard to the dimensions of the TP partition function Q and its relation to Z via the Laplace transform. With the correct Gibbs counting, Eq. (6.10), one obtains, for the TP partition function Q, Eq. (5.6), a quantity with the dimensions of volume, while its logarithm, the Massieu function Ψ_G, should be adimensional. This can be fixed by rescaling Q by some arbitrary constant V_0 with the dimensions of volume. Another possibility is to consider as partition function the normalisation in Eq. (5.3) opportunely rescaled by $h^{3N+1}/N!$ so as to make it adimensional. This amounts to redefine Q as

$$Q(\beta, \Pi, N) = \sqrt{\frac{2\pi M}{h^2 A^2 \beta}} \int dV \int \frac{d\mathbf{z}}{h^{3N} N!} e^{-\Pi V} e^{-\beta H(\mathbf{z}; V)} = \sqrt{\frac{2\pi M}{h^2 A^2 \beta}} \int dV e^{-\Pi V} Z. \tag{6.12}$$

Note how the multiplicative factor within the square root contains the variable β. With the new definition, it is, $\partial \ln Q / \partial \beta = -U - 1/(2\beta)$, $\partial \ln Q / \partial \Pi = -\bar{V}$. For large systems, such that $1/\beta$ is negligible compared to U, the adimensional quantity $\ln Q$ can then be identified with the Massieu potential.

[2] If a gas is made of particles of two distinguishable types, it makes sense both to ask what is the probability of finding N particles in the volume V, regardless of their type, thus arriving at Eq. (6.7), or to ask what is the probability of finding N_1 particles of one type and $N_2 = N - N_1$ particles of the other, in which case, we would arrive, assuming the Poissonian character of both quantities and their statistical independence, to a generalised expression:

From now on we shall adopt the new definitions in Eqs. (6.9 and 6.10), complemented by Eq. (6.11).

Exercise 6.1   Using Stirling formula, $\ln N! \simeq N \ln N - N + O(\ln N)$, show that, for large N the ideal gas canonical free energy $F = -\beta^{-1} \ln Z$, evaluated with the correct counting, obeys the scaling relation $F(\beta, V, N) \simeq N f(\beta, V/N)$, namely it is extensive. Show that the microcanonical entropy, $S = \ln \Omega$, obeys the same scaling.   ■

6.3   Validity of the Heat Theorem within the Grandcanonical Ensemble

Since the grandcanonical distribution is a mixture of canonical distributions with same temperatures, it is not a surprise that the equipartition theorem continues to hold in the grand-canonical ensemble

Theorem 6.1   (Grandcanonical Equipartition Theorem)

$$\left\langle z_i \frac{\partial H}{\partial z_i} \right\rangle_{\beta, V, \alpha} = \frac{1}{\beta} \tag{6.14}$$

where z_i is any of the q's or the p's of the system and $\langle \cdot \rangle_{\beta, V, \alpha}$ denotes average over the grandcanonical distribution.

This tells us that we can assign the meaning of inverse temperature to β.

Exercise 6.2   Prove Theorem 6.1.   ■

We now introduce, as usual, the quantities:

$$T(\beta, V, \alpha) \doteq \left\langle z_i \frac{\partial H}{\partial z_i} \right\rangle_{\beta, V, \alpha} = \frac{1}{\beta}, \tag{6.15}$$

$$P(\beta, V, \alpha) \doteq -\left\langle \frac{\partial H}{\partial V} \right\rangle_{\beta, V, \alpha}, \tag{6.16}$$

$$U(\beta, V, \alpha) \doteq \langle H_N \rangle_{\beta, V, \alpha}, \tag{6.17}$$

$$\bar{N}(\beta, V, \alpha) \doteq \langle N \rangle_{\beta, V, \alpha}. \tag{6.18}$$

Note that N is now fluctuating, hence we have introduced the average number $\bar{N}(\beta, V, \alpha)$. Also, unfortunately, there is no microscopic expression for the chemical

---

$$\rho \, dz_1 dz_2 = \frac{e^{\alpha_1 N_1 - \beta H_{N_1}(z_1; V)}}{\Xi_1(\beta, V, \alpha_1)} \frac{e^{\alpha_2 N_2 - \beta H_{N_2}(z_2; V)}}{\Xi_2(\beta, V, \alpha_2)} \frac{dz_1}{h^{3N_1} N_1!} \frac{dz_2}{h^{3N_2} N_2!}. \tag{6.13}$$

If you can attach a label to each particle, it makes sense to ask for the probability that the particles labelled $i_1 \ldots i_N$ are in the volume, in which case the $N!$ term does not appear.

potential, so we shall assume that is represented by the quantity α/β, see Eq. (1.31). We can now ask ourselves the nontrivial question wether the differential representing the heat differential over temperature

$$\frac{dU(\beta, V, \alpha) + P(\beta, V, \alpha)dV - \mu d\bar{N}(\beta, V, \alpha)}{T(\beta, V, \alpha)}, \tag{6.19}$$

is exact. Not surprisingly, the answer is given by the following

Theorem 6.2 *The differential in Eq. (6.19) is exact, namely there exist a function $S(\beta, V, \alpha)$ such that*

$$dS(\beta, V, \alpha) = \frac{dU(\beta, V, \alpha) + P(\beta, V, \alpha)dV - \mu d\bar{N}}{T(\beta, V, \alpha)}. \tag{6.20}$$

Furthermore, S is given by

$$S(\beta, V, \alpha) = \beta U(\beta, V, \alpha) - \alpha\bar{N}(\beta, V, \alpha) + \ln \Xi(\beta, V, \alpha). \tag{6.21}$$

The proof is based on the following crucial relations:

$$U(\beta, V, \alpha) = -\frac{\partial}{\partial\beta} \ln \Xi(\beta, V, \alpha), \tag{6.22}$$

$$\beta P(\beta, V, \alpha) = \frac{\partial}{\partial V} \ln \Xi(\beta, V, \alpha), \tag{6.23}$$

$$\bar{N} = \frac{\partial}{\partial\alpha} \ln \Xi(\beta, V, \alpha). \tag{6.24}$$

Exercise 6.3 Prove Eqs. (6.22, 6.23 and 6.24) and complete the proof of the theorem. ∎

We note that $\ln \Xi$ is to be identified with the grand Massieu potential:

$$\Psi_\Omega(\beta, V, \alpha) = \ln \Xi(\beta, V, \alpha), \tag{6.25}$$

which is related to the standard grand-potential $\Omega(T, V, \mu)$ by the relation

$$\Omega(T, V, \mu) = -T\Psi_\Omega(1/T, V, \mu/T). \tag{6.26}$$

6.4   Ideal Gas

Let us now come to a most crucial example: an ideal gas. With the correct counting, the canonical partition function reads:

$$Z(\beta, V, N) = \frac{1}{N!} \left[\left(\frac{2\pi m}{h^2 \beta} \right)^{3/2} V \right]^N. \tag{6.27}$$

Let us introduce the thermal wavelength λ (a.k.a., the De Broglie wavelength):

$$\lambda(\beta) = \sqrt{\frac{h^2 \beta}{2\pi m}}. \tag{6.28}$$

The grandcanonical partition function reads

$$\Xi(\beta, V, \alpha) = \sum e^{\alpha N} \frac{1}{N!} \left(\frac{V}{\lambda(\beta)^3} \right)^N = \exp\left(\frac{V e^\alpha}{\lambda^3(\beta)} \right). \tag{6.29}$$

Leading to a very simple expression for the grand-Massieu potential:

$$\Psi_\Omega(\beta, V, \alpha) = \frac{V e^\alpha}{\lambda^3(\beta)}. \tag{6.30}$$

Taking its partial derivatives we obtain:

$$-U = \frac{\partial}{\partial \beta} \ln \Xi(\beta, V, \alpha) = -\frac{3}{2} \frac{\Psi_\Omega}{\beta}, \tag{6.31}$$

$$\beta P = \frac{\partial}{\partial V} \ln \Xi(\beta, V, \alpha) = \frac{\Psi_\Omega}{V}, \tag{6.32}$$

$$\bar{N} = \frac{\partial}{\partial \alpha} \ln \Xi(\beta, V, \alpha) = \Psi_\Omega. \tag{6.33}$$

The second equation establishes the well known expression of the grand potential as the negative PV product [5]

$$PV = -\Omega. \tag{6.34}$$

Combining the first equation with the third one and the second expression with the third one, we obtain the ideal gas thermodynamics:

$$PV = \bar{N}T, \tag{6.35}$$

$$U = \frac{3}{2} \bar{N}T. \tag{6.36}$$

6.4.1 Fluctuations

It is not difficult to see that

$$\frac{\partial^2 \Psi_\Omega}{\partial \alpha^2} = \langle N^2 \rangle - \bar{N}^2, \tag{6.37}$$

where we omit the label β, V, α for simplicity. On the other hand, $\partial \Psi_\Omega / \partial \alpha = \Psi_\Omega$, hence $\partial^2 \Psi_\Omega / \partial \alpha^2 = \partial \Psi_\Omega / \partial \alpha = \Psi_\Omega$. Thus:

$$\langle N^2 \rangle - \bar{N}^2 = \bar{N}, \tag{6.38}$$

which is the typical relation between fluctuations and average value of a Poisson distribution. Note that in the large N limit the relative N fluctuations, $\sqrt{\langle N^2 \rangle - \bar{N}^2}/\bar{N}$ vanish.

The probability distribution of having N particles in the volume V, reads:

$$\mathcal{P}(N; \beta, V, \alpha) = \int \frac{d\mathbf{z}}{h^{3N} N!} \rho(\mathbf{z}, N; \beta, V, \alpha) = \frac{e^{\alpha N} Z(\beta, V, N)}{\Xi(\beta, V, \alpha)}. \tag{6.39}$$

Taking the second derivative of Ψ_Ω with respect to β gives, as in the canonical ensemble, the energy fluctuations:

$$\frac{\partial^2 \Psi_\Omega}{\partial \beta^2} = \langle H^2 \rangle - \langle H \rangle^2 = -\frac{3}{2} \frac{\partial}{\partial \beta} \frac{\Psi_\Omega}{\beta} = \frac{15}{4} \frac{\Psi_\Omega}{\beta^2} = \frac{15}{4} T^2 \bar{N}. \tag{6.40}$$

Note that these fluctuations are larger than the fluctuations with fixed number of particles, which would amount to $3NT^2/2$.

It is interesting to take the second derivative with respect to V. We have:

$$\frac{\partial^2 \Psi_\Omega}{\partial V^2} = \frac{\partial (\beta P)}{\partial V} = \frac{\partial}{\partial V} \frac{\Psi_\Omega}{V} = \frac{\partial}{\partial V} \frac{e^\alpha}{\lambda^3(\beta)} = 0, \tag{6.41}$$

that is

$$\left(\frac{\partial P}{\partial V} \right)_{\beta,\alpha} = 0, \tag{6.42}$$

which means that the compressibility $\chi_{\beta,\alpha} = -V^{-1} \partial V / \partial P|_{\beta,\alpha}$ at fixed temperature and chemical potential, is infinite for an ideal gas in the grand canonical ensemble. To understand this result note that particles are not physically bounded to the region V. Since the box is open, changing the volume of the box while keeping the chemical potential and temperature constant does not correspond to physically compress the system, but merely to a redrawing of an imaginary boundary, hence it will not affect its pressure.

On the other hand, just like in the canonical ensemble, it is:

$$\left\langle \left(\frac{\partial H}{\partial V}\right)^2\right\rangle - \left\langle\frac{\partial H}{\partial V}\right\rangle^2 = \frac{1}{\beta}\left[\frac{\partial P}{\partial V} + \left\langle\frac{\partial^2 H}{\partial V^2}\right\rangle\right], \qquad (6.43)$$

hence the following relation for the fluctuations in the instantaneous force per unit surface on the walls of the open box:

$$\left\langle \left(\frac{\partial H}{\partial V}\right)^2\right\rangle - \left\langle\frac{\partial H}{\partial V}\right\rangle^2 = \frac{1}{\beta}\left\langle\frac{\partial^2 H}{\partial V^2}\right\rangle. \qquad (6.44)$$

The reader can explore independently the value and physical meaning of the cross derivatives.

Exercise 6.4 Study the thermodynamics of the ideal gas in the grandcanonical ensemble without the $N!$ term. That is replace $N!$ with 1 in Eqs. (6.7 and 6.8). (a) Show that the ideal gas laws still holds. (b) Show that $\partial P/\partial V \neq 0$. (c) Write down the explicit expression of the number probability distribution $\mathcal{P}(N)$ and show that the mean root square deviation $\sqrt{\langle N^2\rangle - \langle N\rangle^2}$ is of the same order as $\langle N\rangle$ for large $\langle N\rangle$, namely that the relative fluctuations $\sqrt{\langle N^2\rangle - \langle N\rangle^2}/\langle N\rangle$ are not vanishing in the large $\langle N\rangle$ limit [6]. ∎

Exercise 6.5 Set up a computer program that simulates the dynamics of a gas of $N_{tot} \gg 1$ hard spheres of volume v in a box of volume V_{tot}. Let the total volume occupied by the spheres $N_{tot}v \ll V$. Consider a region of small volume $v \leq V \ll V_{tot}$ contained in the box. Let the system evolve from an arbitrary initial condition and sample e.g., at regular intervals of time, the number of particles N contained the volume V. Build the statistics of N and check whether it conforms to a Poisson distribution. Draw your own conclusions regarding the importance of the $N!$ factorial term in the phase space measure, depending on the results. ∎

Exercise 6.6 Combining Eqs. (6.24 and 6.37) one can express the number fluctuations as $\langle N^2\rangle - \bar{N}^2 = \partial\Psi_\Omega/\partial\alpha$. Using the property of Jacobians show that it can be conveniently recast as (see e.g., [7], Sect. 4.5.1):

$$\langle N^2\rangle - \bar{N}^2 = \beta^{-1}\varrho^2 V\chi_{T,N}, \qquad (6.45)$$

where $\varrho = \bar{N}/V$ denotes the density and $\chi_{T,N}$ is the compressibility at fixed T and N. Note that at variance with $\chi_{\beta,\alpha}$, the quantity $\chi_{T,N}$ is generally not null. ∎

6.5 Remarks

Our approach to the grandcanonical ensemble visibly departs from the way it is usually introduced in statistical mechanics textbooks. Sometimes this ensemble, just like the other ensembles, is simply postulated, without much explanation of its origin. In some cases it is justified based on analogy: Just like the relaxation of constraint on volume leads to the term $-\beta PV$ in the exponent defining the TP distribution, so, analogously, a term $\beta \mu N$ should appear in the exponent when relaxing the constrain on N, see e.g., [8, 9]. A standard derivation is that by Landau [5]. It is implicitly based on Eq. (3.44) for the probability of a subsystem of a larger system, as applied to a quantum system, whereby the eigenenergy replaces the classical Hamiltonian. In this approach the density of states of the reservoir is identified with its exponentiated entropy (i.e., $\omega_R = e^{S_R}$) which is then expanded in Taylor series in the small energy and number of particles of the system (compared to the reservoir). Thermodynamic relations are then used to assign a meaning to the coefficients of the expansion. Despite its simplicity, one might find the standard derivation as not fully satisfactory, from a foundational point of view. The first point to notice is that it assigns the entropy $S_R = \ln \omega_R$ to the reservoir, which is expressed in terms of the structure function ω_R, rather than the phase volume Ω_R of the reservoir. As we shall see in the next chapter, that is not a big problem if the thermodynamic limit exists: in that case the two expressions are equivalent. Another problem is that such expression of entropy, in fact the microcanonical entropy, holds for isolated systems at equilibrium with fixed energy, while, clearly, the reservoir energy is fluctuating. In an effort to avoid these issues, for this lecture, we got inspiration from another classic text namely that of Huang [3]. Huang follows a track that is similar to Landau's, but remains within the realm of classical mechanics, and attempts at establishing a connection with the microscopic dynamics. At close inspection, Huang's derivation is seen to be based on assumptions of spatial homogeneity, large numbers and statistical independence. In short, it implicitly assumes the Poissonian character of the stochastic process governing the statistics of N. Here we have spelled out that assumption, and pointed out that it almost immediately leads to the grandcanonical ensemble, with almost no math and without invoking thermodynamic relations. The only other assumption that we explicitly had to make is Eq. (6.5), which, for sufficiently large N would be justified for systems admitting the thermodynamic limit, namely such that $F(\beta, V, N) = -\beta^{-1} \ln Z \simeq N f(\beta, V)$ which is more or less implicit in other derivations.[3] It is worth stressing, however, that the ensemble in Eq. (6.4), which is parametrised via the inverse temperature β, the volume of the open box V, and the intensity f of the Poisson process, is valid regardless of that assumption.

Hopefully, the present derivation helps the experienced reader gaining further insight about an ensemble that is so central in statistical mechanics, while remaining pedagogical for for those who encounter it here for the first time.

---

[3] We recall that the Z appearing in Eq. (6.5) is the "old" partition function, Eq. (4.11) without the $N!$ correction. With the correct counting, the same condition would read $F(\beta, V, N) = -\beta^{-1} \ln Z \simeq N f(\beta, V/N)$, see Exercise 6.1.

References

1. Larson, R., Odoni, A.: Urban Operations Research. Dynamic Ideas (2007)
2. Kingman, J.: Poisson processes. Oxford Studies in Probability. Clarendon Press (1992)
3. Huang, K.: Statistical Mechanics, 2nd edn. Wiley, New York (1987)
4. Balian, R.: From microphysics to macrophysics. Methods and Applications of Statistical Physics. vol. I. Springer, Berlin, Heidelberg (2007)
5. Landau, L., Lifschitz, E.: Statistical Physics, 2nd edn. Pergamon, Oxford (1969)
6. Fernández-Peralta, A., Toral, R.: Entropy **18**(7), 259 (2016). https://doi.org/10.3390/e18070259
7. Goldenfeld, N.: Lectures on Phase Transitions and the Renormalization Group, 1st edn. CRC Press (1992). https://doi.org/10.1201/9780429493492
8. Feynman, R.P., Leighton, R.B., Sands, M.: The Feynman Lectures on Physics. Addison-Wesley (2006)
9. Peliti, L.: Statistical Mechanics in a Nutshell. In a Nutshell. Princeton University Press (2003)

Chapter 7
Ensemble (in)-Equivalence

7.1 The Problem

In the previous chapters we have considered distinct physical scenarios and have seen how distinct statistical ensembles are appropriate for their respective description. Accordingly, given a certain system, specified by its Hamiltonian, one might obtain different thermodynamic relations among the components of the thermodynamic field, depending on which ensemble is used for their calculation. In this chapter we shall see that typically that doesn't happen, namely, typically ensembles are *equivalent*, and shall comment on whether, when, and in what sense discrepancies are expected, depending on the properties of the system under consideration.

In the following we shall restrict our discussion to the comparison of microcanonical and canonical ensembles. The extension to other ensembles is beyond the scope of these lectures.

Consider the E, V, N (microcanonical) ensemble. The quantity of prime interest is the phase volume $\Omega(E, V, N)$. Its logarithm gives the (microcanonical) entropy $S(E, V, N) = \ln \Omega(E, V, N)$. By Legendre transform one thus obtains the (β, V, N) Massieu potential:

$$\Psi_\mu(\beta, V, N) = \sup_{E}[\ln \Omega(E, V, N) - \beta E], \qquad (7.1)$$

where for simplicity of notation we have dropped the label F, and introduced instead the label μ to stress that it is evaluated in the microcanonical ensemble.

Within the canonical ensemble, the (β, V, N) Massieu potential is directly given by the partition function Z:

$$\Psi_c(\beta, V, N) = \ln Z(\beta, V, N), \qquad (7.2)$$

where we introduced the label c to recall that the evaluation is performed within the canonical ensemble.

© The Author(s), under exclusive license to Springer Nature Switzerland AG 2021
M. Campisi, *Lectures on the Mechanical Foundations of Thermodynamics*,
SpringerBriefs in Physics,
https://doi.org/10.1007/978-3-030-87163-5_7

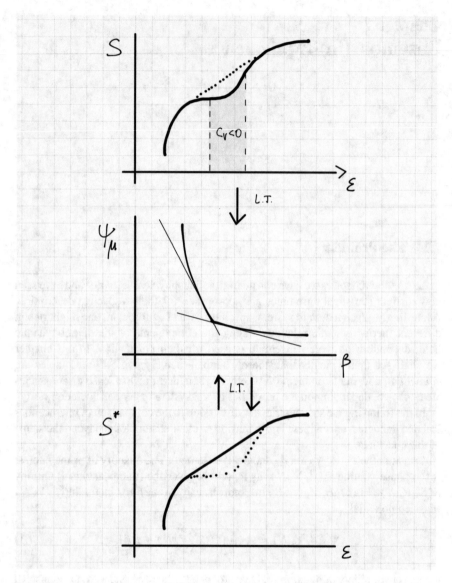

Fig. 7.1 A non concave S, featuring a negative C_V region, is Legendre transformed into a convex Massieu potential Ψ_μ featuring no negative C_V. Further Legendre transform leads to S^*, the concave envelope of S. Both S and S^* have same Legendre transform Ψ_μ

The problem of equivalence is twofold. On one hand it consists in assessing whether the entropy S and its Legendre transform Ψ_μ provide the same thermodynamics. This is an issue that regards the microcanonical ensemble alone. On the other, it consists in assessing whether the two potentials Ψ_c and Ψ_μ coincide, or at least if their partial derivatives coincide, in the large N limit.

The first aspect might seem trivial at first sight. The Legendre transforms are introduced exactly for the propose of changing the variables without changing the equations of state. The subtle point is that this works well as long as the function to be transformed is either strictly concave or strictly convex. If that is not the case some information is lost in the process of transforming and equivalence of the two descriptions is lost. To see that let us recall that because of the general properties of the Legendre transforms, Ψ_μ is a convex function, regardless of the concavity of S. Hence while S might predict negative heat capacity,

$$C_V = -\frac{(\partial S/\partial E)^2}{\partial^2 S/\partial E^2},\tag{7.3}$$

in the region of convexity $\partial^2 S/\partial E^2 > 0$, the potential Ψ_μ may only predict non-negative $C_V = \beta^2 \partial^2 \Psi_\mu/\partial\beta^2$ because everywhere it is $\partial^2 \Psi_\mu/\partial\beta^2 > 0$.[1] From the mathematical point of view that is reflected in the fact that if S is concave everywhere, then S^* (its double Legendre transform) coincides with S itself. Instead, if S has some regions where the concavity is inverted, then S^* which by definition is concave (because it is the transform of a convex function Ψ_μ), clearly does not coincide with S (in fact S^* is the concave envelope of S), thus highlighting that some information gets lost in the process of transforming S, (Fig. 7.1).

In the following we shall consider a number of prototypical illustrative situations showing various degrees of equivalence, listed from higher to lower.

7.2 Full Equivalence

We consider the case of the ideal gas. The phase volume, including the correct Gibbs counting reads,

$$\Omega(E, V, N) = \frac{K^{3N/2}}{N!\Gamma(1 + 3N/2)} V^N E^{3N/2},\tag{7.4}$$

where $K = 2m\pi/h^2$. $S = \ln\Omega$ is strictly concave everywhere. Taking the logarithm and then performing the Legendre transform one arrives at:

---

[1] This is in fact true for Ψ_c as well, see Eq. (4.35).

$$\Psi_\mu(\beta, V, N) = \frac{3N}{2} \ln \beta^{-1} + N \ln V - \ln N! + \frac{3N}{2} \ln K$$
$$+ \frac{3N}{2} \ln \frac{3N}{2} - \frac{3N}{2} - \ln \Gamma\left(1 + \frac{3N}{2}\right). \qquad (7.5)$$

On the other hand, from the canonical partition function

$$Z(\beta, V, N) = \frac{K^{3N/2}}{N!} V^N \beta^{-3N/2}, \qquad (7.6)$$

we obtain

$$\Psi_c(\beta, V, N) = \frac{3N}{2} \ln \beta^{-1} + N \ln V - \ln N! + \frac{3N}{2} \ln K. \qquad (7.7)$$

Therefore:

$$\Psi_\mu(\beta, V, N) = \Psi_c(\beta, V, N) + \frac{3N}{2} \ln \frac{3N}{2} - \frac{3N}{2} - \ln \Gamma\left(1 + \frac{3N}{2}\right). \qquad (7.8)$$

We note that regardless of the value of N, the functional dependence of the two potentials on β and V are identical. Accordingly, one gets the same state equations in both ensembles even without taking the large N limit. This is a very special case. As we shall prove below, typically, equivalence holds in the thermodynamic limit only.

We further note that, in the large N limit, using the Stirling formula $\Gamma(1 + x) \simeq x \ln x - x$, we have

$$\Psi_\mu(\beta, V, N) \simeq_{N \gg 1} \Psi_c(\beta, V, N), \qquad (7.9)$$

that is, for increasing N the two potentials tend to coincide in value.

7.3   Equivalence

The above example consisting of a system of noninteracting particles, where equivalence holds already for finite N is very special. As soon as more realistic cases of many-body interacting systems are considered the degree of equivalence drops down.

7.3.1   Hilbert and Dunkel Model of Evaporation

To exemplify that we consider a 1D model of an interacting gas that has been solved exactly few years ago by Hilbert and Dunkel [1]. It consists of N particles confined in a box of length L, see Fig. 7.2:

Fig. 7.2 The model of
Hilbert and Dunkel. N disks
are confined to a line
segment of length L and
interact via a square potential
featuring hard core repulsion
and short range attraction

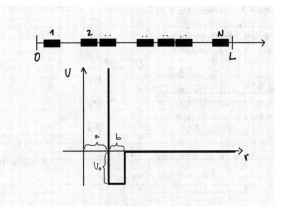

$$U_{\text{box}}(\mathbf{q}) = \begin{cases} 0 & \mathbf{q} \in [0, L]^N \\ +\infty & \text{otherwise} \end{cases}. \tag{7.10}$$

The particles interact via an idealised square well potential featuring a repulsive hard-core and short range attraction:

$$U(r) = \begin{cases} +\infty, & r \leq a \\ -U_0, & a < r < a+b \\ 0, & r \geq a+b \end{cases}. \tag{7.11}$$

Here $a > 0$ is the radius of the hard-core, $b > 0$ is the interaction range. This interaction potential is an idealisation of the Lennard-Jones type potential that well describes many intermolecular interactions. Under the conditions:

$$b \leq a, \tag{7.12}$$

$$L > (N-1)(a+b), \tag{7.13}$$

ensuring that each particle interacts only with its nearest neighbours, and that the fully dissociated state of potential energy $E = 0$ is possible; the phase volume takes the form:

$$\Omega(E, L, N) = C(N) \sum_{k=0}^{N-1} \omega_k(L, N)[E + kU_0]_+^{N/2}, \tag{7.14}$$

where $[x]_+^p = x^p \theta(x)$, with θ the Heviside step function, and

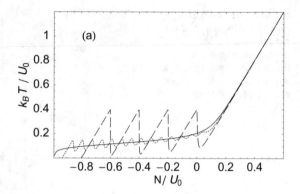

Fig. 7.3 Temperature T as functions of energy per particle ϵ in the model of Hilbert and Dunkel, for various values of $N = 5, 15, 500$. Oscillations increase in number and decrease in amplitude as N grows. Reprinted with permission from [1]. Copyright (2006) by the American Physical Society

$$C(N) = \frac{2(2\pi m)^{N/2}}{N!\Gamma(N/2 + 1)h^N}, \tag{7.15}$$

$$\omega_k(L, N) = \binom{N-1}{k} \sum_{i=0}^{k} \binom{k}{i}(-1)^i[L - (n-1)a - b(N-1-k+i)]^N, \tag{7.16}$$

where $\binom{i}{j}$ is the binomial coefficient. The interested reader can find the derivation in the Appendix of Ref. [1].

By Laplace transform one obtains the canonical partition function, and canonical Massieu potential:

$$Z(\beta, L, N) = \frac{1}{N!}\left(\frac{2\pi m}{\beta h^2}\right)^{N/2} \sum_{k=0}^{N-1} \omega_k(L, N)e^{k\beta U_0}, \tag{7.17}$$

$$\Psi_c(\beta, L, N) = \ln Z(\beta, L, N). \tag{7.18}$$

It is crucial to note that $S = \ln \Omega$ is not concave. In fact the function presents $N - 1$ *convex intruders*. Those are evident when plotting the temperature $T = (\partial S/\partial E)^{-1}$, showing $N - 1$ oscillations, see Fig. 7.3. The regions where $\partial T/\partial E$ is negative signal the region of negative heat capacity, namely of convexity of the entropy. In those regions there cannot be equivalence with the canonical ensemble, where by construction the heat capacity is a positive quantity. Note also that because of the convex intruders there is no unique solution to the equation

$$\frac{\partial S}{\partial E} = \beta, \tag{7.19}$$

that is, there is no one-to-one correspondence between E and β. However, in the limit of large N, the amplitude of the temperature oscillations becomes smaller and smaller, the convex intruders accordingly vanish, and Ψ_c goes over Ψ_μ, thus establishing equivalence.

As a side note we mention that the function Ω has $N - 1$ non-analytic points at energies $E_k = -kU_0$, whose order of nonanaliticity grows indefinitely with N, hence the entropy S becomes smooth in the large N limit. What remains, after taking the large N limit is a region of small slope of the caloric curve, hence of large C_V. In dimension $d > 1$, one expects that slope to go exactly to zero, i.e., $C_V = \infty$, which is the hallmark of a first order phase transition, see Sect. 7.4 below.

7.3.2 The Saddle Point Approximation

The large N equivalence observed in the Hilbert-Dunkel model is rather general, and occurs any time there is a well defined thermodynamic limit, namely whenever in the limit of large N and V, it is:

$$S(E, N, V) \to Ns(\epsilon, \nu), \tag{7.20}$$

where $\epsilon = E/N$, $\nu = V/N$, and s is a strictly concave function, as in the above case.

In fact (adopting the convention that the minimum of energy is $E = 0$)

$$Z(\beta, V, N) = \beta \int_0^\infty dE\, \Omega(E, V, N) e^{-\beta E} = \beta \int_0^\infty dE\, e^{S(E,V,N)-\beta E}$$

$$= N\beta \int_0^\infty d\epsilon\, e^{N[s(\epsilon,\nu)-\beta\epsilon]}, \tag{7.21}$$

since both S and E scale with N, the integral is dominated by the value of E where the exponent is maximal. Let $\bar{E}$ be such value, namely let it be the (unique) solution of

$$\frac{\partial S}{\partial E} = \beta, \tag{7.22}$$

Expanding the exponent around $\bar{E}$:

$$S - \beta E = \sup_E [S - \beta E] + \frac{1}{2}(E - \bar{E})^2 \frac{\partial^2 S}{\partial E^2} + \cdots$$

$$= \Psi_\mu(\beta, V, N) - \frac{\beta^2 (E - \bar{E})^2}{2C_V} + \cdots \tag{7.23}$$

we obtain:

$$Z(\beta, V, N) \simeq e^{\Psi_\mu} \beta \int_0^\infty dE\, e^{-\frac{\beta^2 (E-\bar{E})^2}{2C_V}} \simeq e^{\Psi_\mu} \beta \sqrt{2\pi\beta^2 C_V}, \tag{7.24}$$

hence:

$$\Psi_c \simeq \Psi_\mu + \ln\left(\beta^2\sqrt{2\pi C_V}\right). \tag{7.25}$$

Note that while Ψ_c and Ψ_μ scale with N, their difference is of order $\ln N$, hence can be neglected in the large N limit, thus showing that canonical and microcanonical Massieu potentials coincide in this limit.

7.4 Partial Equivalence

When considering more realistic cases, for example a 3D system with short-range Lennard-Jones type interactions among the particles, the rescaled entropy $s(\epsilon, \nu)$ may present a region, say the region $[\epsilon_1, \epsilon_2]$, of constant slope β_t, namely where $\partial s/\partial \epsilon$ is constant, that is C_V, the heat capacity diverges, see Fig. 7.4. That is associated with a first order phase transition signaling that a small addition of energy is used up by the system to break molecular bonds, rather than for increasing the kinetic energy, hence the temperature. This gives origin to the latent heat associated to the liquid-gas phase transition. At the transition inverse temperature β_t the function $e^{S-\beta E}$ presents a plateaux and is of order e^N. The integral is dominated by the plateaux region, bringing roughly the contribution $e^{\Psi_\mu} N(\epsilon_2 - \epsilon_1)$, which yet is dominated by e^{Ψ_μ} for large N. Thus the canonical and microcanonical Massieu potentials continue to coincide in the large N limit.

The inequivalence is only marginal in the sense that the one-to-one correspondence of states in the two ensembles is valid everywhere except at one single value β_t which is associated to all the values of energy in the range $[\epsilon_1, \epsilon_2]$.

7.5 Inequivalence

If long-range interactions are present, it may happen that a convex intruder in the microcanonical entropy survives even in the large N limit. If that happens there will be an extended region where the microcanonical description predicts a negative heat capacity, whereas the canonical one will predict a non-negative one, see Fig. 7.5. The two descriptions are inequivalent, regardless of whether Ψ_μ and Ψ_c would coincide or not.

Physical examples of the present situation include self-gravitating astrophysical objects, e.g., stars, which are thermally insulated (not immersed in a bath), and dominated by gravitation which is a long range force. In certain cases, addition of energy results in the star cooling down, that is the heat capacity is negative. Such systems cannot evidently be correctly described by the canonical formalism.

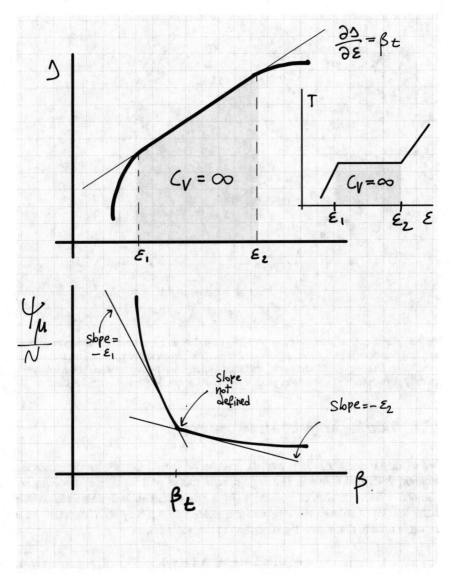

Fig. 7.4 Sketch of entropy and Massieu potential in presence of a first order phase transition. The slope of s (hence the temperature T) is constant in the range $[\epsilon_1, \epsilon_2]$, resulting in a diverging heat capacity C_V associated to the latent heat. The Massieu potential is non-analytic at the critical inverse temperature β_t, where left and right derivatives do not coincide. The slope at β_t, representing internal energy, is not defined. This corresponds to the fact that all possible mixtures of coexisting phases, whose energy density lies between ϵ_1 and ϵ_2, have same temperature. The inset shows the according caloric curve

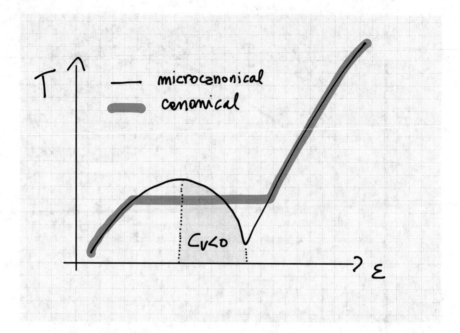

Fig. 7.5 Sketch of typical caloric curve for a long range interacting system. The convex intruder of the entropy survives in the limit of large N, resulting in an extended region where the microcanonical ensemble predicts negative heat capacity, that cannot be captured by the canonical ensemble, which typically predicts instead a first order phase transition. The two description are not equivalent

7.5.1 Thirring Artificial Model of a Star

One of the simplest idealised models that can be solved exactly, and presents the situation described above is Thirring artificial model of a star [2]. N particles are confined within a region of volume V by a box potential. Inside this region is a subregion $\mathcal{V}_0$ of volume V_0. Two particles interact only if they both happen to be at the same time in the region V_0, according to the potential:

$$\phi(\mathbf{x}_i, \mathbf{x}_j) = -v\chi(\mathbf{x}_i)\chi(\mathbf{x}_j), \tag{7.26}$$

where

$$\chi(\mathbf{x}) = \begin{cases} 1, & \mathbf{x} \in \mathcal{V}_0 \\ 0, & \mathbf{x} \notin \mathcal{V}_0 \end{cases}. \tag{7.27}$$

is the characteristic function of the set $\mathcal{V}_0$, and v is some positive constant, ensuring that the interaction is attractive. The phase volume can be calculated explicitly as [2]:

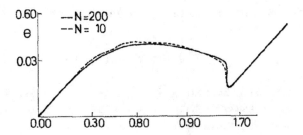

Fig. 7.6 The caloric curve $\theta(\varepsilon)$ of Thirring artificial model of a star for $N = 10, 200$ and $F \doteq \ln(V/V_0) = 4.5$. Reprinted by permission from Springer Nature Customer Service Centre GmbH: Springer Nature, Ref. [2], Copyright (1970)

$$\Omega(E, V, N) = \frac{V^N \pi^{3N/2}}{(3N/2)!} \sum_{k=0}^{N} \frac{(E + k^2 v)^{3N/2} V^{N-k}}{k!(N-k)!}. \tag{7.28}$$

Figure 7.6 shows typical plots of the dependence of the rescaled temperature θ as a function of rescaled energy ε, which are defined as

$$E = N^2 v(\varepsilon - 1), \quad T = \frac{2}{3} N v \theta. \tag{7.29}$$

in accordance with the fact that in this model the energy scales quadratically in energy, and the temperature scales linearly [2].

Exercise 7.1 (a) Derive Eq. (7.28). (b) Use a software of your choice to reproduce the plots in Fig. 7.6. (c) Use a software of your choice to plot the canonical caloric curve. How does it compare to the microcanonical one? ∎

7.6 (In)-equivalent Expressions of the Microcanonical Entropy

We remark that in most statistical mechanics textbooks the microcanonical entropy is not derived from mechanics, as in the present book, but is postulated to be given by the logarithm of the density of states:

$$S_\omega(E, V, N) = \ln \omega(E, V, N). \tag{7.30}$$

It is important to remark that, generally, ω is not an adiabatic invariant, nor is it the generator of the differential $(dE + \Lambda d\lambda)/T$. Its employment is rather rooted in probability theory [3]. Note also that ω has the dimension of an inverse energy, hence the argument of the logarithm is not dimensionless, meaning that one must introduce as well an arbitrary energy scale ε that renormalises ω in order to obtain a proper definition:

$$S_{\omega,\varepsilon}(E, V, N) = \ln[\omega(E, V, N)]/\varepsilon. \tag{7.31}$$

When the thermodynamic limit exists, $S_{\omega,\varepsilon}$ becomes equivalent to S in the large N limit.[2] To see that, consider the temperature $T_{\omega,\varepsilon}$, and generalised force $\Lambda_{\omega,\varepsilon}$ that one obtains by using $S_{\omega,\varepsilon}$ instead of S[3]:

$$T_{\omega,\varepsilon} = \left(\frac{\partial S_{\omega,\varepsilon}}{\partial E}\right)^{-1}, \tag{7.32}$$

$$\Lambda_{\omega,\varepsilon} = \frac{\partial S_{\omega,\varepsilon}}{\partial \lambda}\left(\frac{\partial S_{\omega,\varepsilon}}{\partial E}\right)^{-1}. \tag{7.33}$$

Using the relation $\omega = \partial\Omega/\partial E$ it is not difficult to prove that they are related to T and Λ via the equations:

$$T_{\omega,\varepsilon} = \frac{T}{1 - C_V^{-1}}, \tag{7.34}$$

$$\Lambda_{\omega,\varepsilon} = \Lambda + T_{\omega,\varepsilon}\frac{\partial\Lambda}{\partial E}. \tag{7.35}$$

If for large N, $S = Ns(E/N, \lambda/N)$, then $\partial S/\partial E = s_1$,[4] and $\partial^2 S/\partial E^2 = s_{11}/N$, hence $C_V = -\frac{(\partial S/\partial E)^2}{\partial^2 S/\partial E^2} = -Ns_1^2/s_{11}$ scales with N. Similarly $\Lambda = \frac{\partial S}{\partial \lambda}/\left(\frac{\partial S}{\partial E}\right)^{-1} = s_2/s_1$, hence $\partial\Lambda/\partial E = (s_{21}s_1 - s_2s_{11})/(Ns_1^2)$ scales with $1/N$. Therefore, when the thermodynamic limit exists

$$T_{\omega,\varepsilon} \simeq T, \tag{7.36}$$
$$\Lambda_{\omega,\varepsilon} \simeq \Lambda. \tag{7.37}$$

where the corrections are of order $1/N$.

Exercise 7.2 Prove Eqs. (7.34 and 7.35). ∎

Exercise 7.3 Show that Eq. (7.36) continues to hold when the thermodynamic limit is obeyed in the generalised form $S \simeq N^\alpha s(E/N^\beta)$ for some positive α, and quantify the order of the correction term. Note that for the Thirring model it is $\alpha = 1$, $\beta = 2$. ∎

---

[2] It must be remarked that from a practical point of view, the evaluation of Ω is more straightforward than that of ω.

[3] Note that these quantities do not depend on ϵ.

[4] The symbol $s_{ij..}$ stands for the partial derivative of s with respect to its i-th argument, followed by the partial derivative with respect to its j-th argument, and so on.

Exercise 7.4 How do the equations of state of an ideal gas change when replacing S with $S_{\omega,\varepsilon}$? Check that equivalence is restored in the large N limit. ∎

References

1. Hilbert, S., Dunkel, J.: Phys. Rev. E **74**, 011120 (2006). https://doi.org/10.1103/PhysRevE.74.011120
2. Thirring, W.Z.: Phys. B **235**, 339 (1970). https://doi.org/10.1007/BF01403177
3. Boltzmann, L.: Sitzungberichte der Kaiserlichen Akademie der Wissenschaften. Mathematisch-Naturwissen Classe. Abt. II, LXXVI (1877), pp 373–435 (Wien. Ber. 1877, 76:373–435). Reprinted in Wiss. Abhandlungen, Vol. II, reprint 42, pp 164–223, Barth, Leipzig, 1909. Translated in English by K. Sharp and F. Matschinsky, Entropy **17**(4), 1971–2009 (2015). https://doi.org/10.3390/e17041971

Printed in the United States
by Baker & Taylor Publisher Services